A General Systems Philosophy for the Social and Behavioral Sciences

THE INTERNATIONAL LIBRARY OF
SYSTEMS THEORY AND PHILOSOPHY
Edited by Ervin Laszlo

THE SYSTEMS VIEW OF THE WORLD
*The Natural Philosophy of the New
Developments in the Sciences*
by Ervin Laszlo

GENERAL SYSTEM THEORY
Foundations—Development—Applications
by Ludwig von Bertalanffy

ROBOTS, MEN AND MINDS
Psychology in the Modern World
by Ludwig von Bertalanffy

THE RELEVANCE OF GENERAL SYSTEMS THEORY
*Papers Presented to Ludwig von Bertalanffy
on His Seventieth Birthday*
Edited by Ervin Laszlo

HIERARCHY THEORY
The Challenge of Complex Systems
Edited by H. H. Pattee

THE WORLD SYSTEM
Models, Norms, Applications
Edited by Ervin Laszlo

A GENERAL SYSTEMS PHILOSOPHY
FOR THE SOCIAL AND BEHAVIORAL SCIENCES
by John W. Sutherland

Further volumes in preparation

A GENERAL SYSTEMS PHILOSOPHY FOR THE SOCIAL AND BEHAVIORAL SCIENCES

JOHN W. SUTHERLAND

GEORGE BRAZILLER   *New York*

Standard Book Number: 0-8076-0724-X, cloth
0-8076-0725-8, paper
Library of Congress Catalog Card Number: 73-86261
First Printing
Printed in the United States of America

Whatever withdraws us from the power of the senses; whatever makes the past, the distant, or the future, predominate over the present, advances us in the dignity of thinking beings.

SAMUEL JOHNSON

Preface

This book is for my colleagues in the social and behavioral sciences, especially those who are curious about general systems theory's pretentions as a new epistemology or innovative methodological platform. It is also for those systems theorists who wonder about the ways in which their constructs and concepts might specifically serve the causes of the human sciences: economics, sociology, political science, social psychology, organization theory, anthropology, etc. The need for such a book is fairly clear-cut. Scarcely any serious social or behavioral scientist can be found who has not, at some point in his career, at least casually questioned the very foundations of his discipline. The general systems theorist, on the other hand, makes his primary mark by constantly questioning the methods and intentions of science. In effect, though he may belong formally to any of several dozen substantive disciplines, his first attention must be to the epistemological predicates of science in general. This is so because general systems theory is not really a theory at all—it is a fundamentally new approach to scientific analysis, an approach which stands in both logical and procedural opposition to more traditional schemas such as strict empiricism, positivism, intuitionalism, or phenomenology. True, it draws its precepts eclectically from all these, but in the process of selection becomes something very different than its components.

Just why it is different, and the many implications of the points of difference, will occupy most of our time here. But there is also a second and constant ambition for this volume: to try to express the properties of general systems theory in ways which make its functionality explicitly clear to the practicing social or behavioral scientist, or to students of the social and behavioral sciences. It is definitely true that this work seeks to make converts to the general systems theory camp. But it is not true that the perspective is an evangelistic one (or simply an exegetical exercise aimed at collecting pregnant quotations from authors sympathetic to the cause). Rather, the claims which general systems theory makes on the epistemological level are analyzed logically and, where any evidence exists, empirically. The epistemological implications must then be translated into points of procedural or methodological significance. This, obviously, is where the test for the individual reader must be made: If the practical significances are appealing, then perhaps a shift in one's personal investigatory approach is indicated, at least on a tentative basis. So, the realistic ambition is simply to make the reader aware that an alternative scientific platform exists, and hopefully make him sufficiently curious about its potential to give it a try.

On another and ancillary dimension, the work here will also presume to speak to the systems theorist himself. As many probably already know, there is great confusion about what is and what is not systems theory, and how general systems theory differs from such special system theories as cybernetics, process control, system engineering, etc. Part of the problem stems from the fact that we, as advocates of *the* systems approach, have largely failed to develop a coherent statement of its properties. True, the field is an emerging rather than mature one, and is far from final in its own configuration. But, even at this point of change and conceptual flux, any effort to treat systems theory as a 'system' in its own right is not to be discouraged. So, to a certain extent, what is attempted here is the presentation of the predicates of general systems theory in a hopefully systemic and nonelliptical scheme, done entirely in the spirit of that old admonition: *Physician, heal thyself!*

Overall, then, what we will be doing in these pages is neither particularly erudite nor revolutionary. In fact, in the most basic sense, it is simply a response to the already very-well developed and documented case for science to examine itself, especially in the light of the considerable disenchantment we must feel about our failure to solve problems in the social and behavioral domain while our technological triumphs keep marching on. Thus, while what general systems theory has to say is by no means restricted to the domain of the human sciences, this appears the most appropriate place to start—if not for ourselves, at least for the sake of those our disciplines exist to serve (and who are now beginning to ask us for so much more of a contribution than we have been able to give in the past).

As a final note, some thanks are in order. First, a note of appreciation to a constant and patient mentor, Eric Trist. Secondly, I want to make it clear that the major debt for any of the really insightful points which will be found here goes to the late Ludwig von Bertalanffy—indeed, almost everyone working in the general systems field would have to make a similar assignment. Lastly, a word of special gratitude to the supportive, kind and very perspicacious editor of this series, Ervin Laszlo; that there is this book at all owes so much to his efforts and suggestions.

New Brunswick, N.J.
May, 1973

Contents

Introduction

An eminent philosopher of science, Errol Harris, has cast a rather disparaging eye at the state of the modern social and behavioral sciences:

The salient characteristic of our time is the conspicuous contrast between the achievements of the human intellect in science and technics and its abysmal failures in the spheres of morals and politics. Yet both of these spheres of activity have traditionally been regarded as the products of reason. . . . there is a sense of the word 'reason' in which both theoretical science and social order are its products. If this is so we are faced with the paradox that the same human intellectual capacity has produced, in one sphere, the most spectacular results, and, in the other, only the most dismal failures.[1]

There is one defense which we, as social and behavioral scientists, can make against such allegations: the point that the subjects with which we as social and behavioral scientists have elected to work are inherently more complex than those which the natural and physical sciences have elected. Yet this apology for the state of our arts remains something of a speculation, for we have as yet evolved no formal mechanism for comparing the inherent complexity of one phenomenon with that of another. In the course of this little volume, then, we shall have to pay detailed attention to this problem of comparative phenomenal complexity. It is both interesting in itself, and critical to a system perspective on the social and behavioral sciences as phenomena in their own right.

But a far more promising avenue for research into the differential progress made in the natural-physical and social-behavioral sciences rests in another domain entirely: the epistemological. It is here that we are concerned about the fundamental perspective which a scientist brings to his field, which in turn determines the subjects he will elect to study, the methodological procedures he will employ and therefore the character of the results he obtains. And it is on this dimension that we shall expect to find the most critical points of difference between the two broad divisions of scientific enterprise. Particularly, we shall note the lack of articulation between theory and practice in the social and behavioral sciences, a startling symptom of the dichotomization of our disciplines into two effectively polarized methodological camps: the empiricist-positivist on the one hand and, on the other, the deductivist-idiographic. Moreover, not only is there a lack of articulation within the disciplines of the social and behavioral sciences, but also *between* the disciplines. That is, we have partitioned our science into parochialized segments even though the phenomena we collectively study are not so segmented. In short, it is highly unlikely that we could strike any meaningfully, properly defined empirical (i.e., real-world subject) whose behavior was solely determined by variables drawn from just a single discipline.

There are two other problems which we shall encounter on the epistemological domain, both of them serious and somewhat embarrassing: First, the tendency to foster competitive theories, each purporting to explain the behavior of the same phenomenon, and then to become evangelists for one or another of these warring grand concepts, often without troubling to examine the premises or predicates of the constructs as to their empirical validation (or lack of it). To this extent, we often become more interested in changing society than in understanding it. Moreover, we tend to attenuate this parallelism rather than resolve it—in short, there may be several psychologies, several economics, several anthropologies, each as fundamentally different in their premises as were the platforms of our progenitors: Plato and Aristotle in their roles as sociobehavioral theorists. Sadly, our most fundamental social, political, and ontological platforms within the modern social

4

and behavioral sciences illustrate that there has been little success in forcing convergences between a priori polarized stances in the intervening millenia. We have psychology divided into Skinnerean and Maslowian camps, to adopt a somewhat crude perspective; anthropologists tend, at the very beginning of their careers, to become either functionalists or evolutionists, each having only minimal points in common with the other; sociologists adopt either the mechanical or organic ideal-type, and thereafter act mainly as apostles for their respective creeds; administrative scientists tend to explain phenomena either in terms of quasi-mathematics and applied microeconomics, or in terms of some variation on the Platonic-Rousseauvian model; and in economics, in broad effect, we tend to revert to one of two essentially polaric (and historical) platforms: the Smithian 'invisible hand' or some modern interpretation of the Platonic centralistic theme. As such, what constitutes any discipline within the social and behavioral sciences depends, largely, on who you happen to be talking to at the moment and on what particular 'school' he attends to.

The second problem lies in the nature of the models we construct as social or behavioral scientists, compared to those constructed in the physical-natural sciences. Particularly, those of the latter will tend to be Janus-faced: one face pointing toward the empirical justifications for its existence, the other face maintaining communication with the more encompassing constructs in the hierarchy of their science. Our models, on the other hand, tend to be relatively insular, pointing out neither the requisites for empirical justification nor the broader conceptual roots which serve as the model's context-independent rationale (if, indeed, there are any historical or synchronic connections). In short, the models, paradigms, and theories of the progressive, concatenative sciences do not spring from the scientist *ab initio* and *ab intra*. Rather, they are explicitly articulated both with the conceptual predicates of

5

their field and with the emerged (or emergent) empirical data bases which will serve either to validate or invalidate them— at least partially. By the nature of their construction, however, the models of the social and behavioral sciences tend to live a life apart, and as such tend to become ephemerals of emotive significance rather than connectives of nomothetic or universal significance.

We mention these problems not by way of recrimination or out-of-hand criticism, but to provide some foci for the epistemological-procedural platform we want to introduce: that of general systems theory. Initially, it presents a methodological approach designed especially to deal with phenomenal complexity—and few will deny that the subjects with which we are forced to deal are sufficiently complex to make this of interest. Secondly, general systems theory can serve as a distinct epistemological alternative to the platforms which currently predominate in the social and behavioral sciences (and to which we can attribute the primary cause for the problems we outlined): strict empiricism, positivism, and phenomenology at one extreme of the continuum and, toward the other, intuitionalism, subjectivism, and rhetoric-idiographic preferences.

But in the present book we shall not be acting as an out-of-hand evangelist for general systems theory. Rather, we shall be exploring its potential within our disciplines on four dimensions: (1). Its promise to provide greater *rectitude* than existing epistemologies, especially in its demand that we adopt a triadic ontology; (2). Its potential for *efficiency*, largely in its concentration on the provision of conceptual envelopes which act as integrative, encompassing agents directed at introducing 'economy of thought'; (3). Its promise to invoke greater *relevance* in the outputs of the social and behavioral sciences, stemming primarily from its demand for a holistic approach to our subjects and its dictate for interdisciplinary attack; (4). Finally, its ability to lend *substantive insights* which are unavailable from any other perspective, especially those which are a joint product of conceptualization and empiricalization linked together in an action-research process.

Now, nothing comes for free, especially in the scientific province. To reap these benefits promised by general systems theory, we are also going to have to make some sacrifices. Particularly, we are going to have to be prepared to look beyond the comfort and security of neat disciplinary niches, to abandon the crossword-puzzle type amusement of simple hypothesis-testing and the quick completion-experiences such activities reward us with. And, most fundamentally, we are certainly going to have to abandon the arrogance, detachment, and insularity which strict empiricism allows the scientist and, in its stead, take responsibility for the consequences of our actions on the communities we ostensibly exist to serve. On the other side of the coin, the grand theorist, so long allowed his effective isolation from the operational business of his discipline and from the implications of experience for his speculations, will have to be concerned with two dictates: That his conceptual devices be open to empirical validation on the one hand and, on the other, that they articulate explicitly with preexisting or higher-level constructs of the social and behavioral sciences. In short, neither the license of the poet nor that of the alchemist is to be granted the scientist operating under the banner of general systems theory. In concept, at least, it would be hard to find a better recommendation for an epistemological platform than its denial of just these licenses.

The adoption of the general systems theory mantle, then, carries both a responsibility and a recommitment—the former to the enterprise of science itself, the latter to the society we ostensibly exist to serve. And the task of the general systems theorist, as Ervin Laszlo has so correctly pointed out:

. . . is not one for the cautious scholar who insists on making sure that nothing he says can be disproved or challenged. But then, these are not times when cautious scholarship can help us set our

7

sights. The processes of change are too explosive and the time too short.[2]

Perhaps, considering the social or behavioral scientist in the light of the physician—committed jointly to truth and to the patient in pain—we might find guidance in this Hippocratic aphorism spun so long ago: *The life so short; the craft so long to learn; the occasion so urgent.*

In the broadest sense, general systems theory may be portrayed as one of the fundamental epistemological platforms available to the modern scientific community. When its ontological and methodological significances are properly articulated, orchestrated with an eye toward specific scientific prescriptions, we emerge with an array of propositions which stands as a generic antithesis to the theses of naive universalism, strict positivism, reductionism, intuitionalism, and phenomenology. In short, the scientist operating under the dictates of the general systems theory platform brings to his subjects a very different perspective—and an essentially different mode of inquiry—than his counterparts acting under traditional epistemological envelopes.

Writing from the perspective of the philosopher, Ervin Laszlo cites the uniqueness of the general systems theory approach:

In the history of Western science, atomistic and holistic ways of thinking have alternated. Early scientific thinking was holistic but speculative; the modern scientific temper reacted by being empirical but atomistic. Neither is free from error, the former because it replaces factual inquiry with faith and insight, and the latter because it sacrifices coherence at the altar of facticity. We witness today another shift in ways of thinking: the shift toward rigorous but holistic theories. This means thinking in terms of facts and events in the context of wholes, forming integrated sets with their own properties and relationships. Looking at the world in terms of such sets of integrated relations constitutes the systems view. It is the present and next choice over atomism, mechanism, and uncoordinated speculation.[3]

If we ask who abides by this next and present choice, in terms of the scientific perspective it lends, we find scholars distributed across virtually all scientific fields: In biology there is Paul Weiss, Robert Rosen, Howard Pattee, James Miller, and the late Ludwig von Bertalanffy, perhaps the man most responsible for the articulation of general systems theory; in anthropology there is Leslie White; in psychology there are many names, predominant among them Gordon Allport, Piaget, Maslow, J. Bruner; in American psychiatry there are influences to be found in the work of Menninger, Arieti, Grinker, and most especially Rizzo and Gray; in linguistics, David McNeil and Noam Chomsky represent tendencies toward the general systems perspective; in engineering there is the work of George Klir, Zadeh and Polak; in economics there is Kenneth Boulding, one of the original founders of the Society for General Systems Research; in the behavioral sciences we can count Eric Trist, Arthur Koestler, Daniel Katz, and Robert Kahn and the late Kenneth Berrien; in sociology there is a systems overtone to the work of Merton and Sorokin and specific system underpinning in the studies of Walter Buckley; in the policy sciences and the management sciences there is Ackoff, Churchman, Fred Emery, Julius Stulman, Lasswell, Simon, Easton and Geoffrey Vickers; in philosophy and ethics there is the distinct work of Ervin Laszlo and the occasional systems slant taken by Tillich; there are hierarchicalists such as the late L. L. Whyte and the Wilsons; historians and philosophers of science, particularly Errol Harris and Thomas Kuhn, either deliberately or inadvertently contribute to the reification of a general systems theory perspective, as do great generalists such as Mesarovič and interdisciplinarians such as Anatol Rapoport, Waddington, Norbert Wiener, and Dubos—and this does not by any means complete the list.

There are, moreover, at least three journals which explicitly seek out and publish contributions from the general systems theory perspective to the social and behavioral sciences. The journal *Behavioral Science*, now officially associated with the Society for General Systems Research, is one such, along with *Human Relations*, and the *International Journal of General Systems*. In addition, especially for the mathematical sophisticate, there is the journal of *Mathematical Systems Theory*; and for the engineer, the publications of the I.E.E.E. on systems, cybernetics, and control, etc. And there is the more eclectic *International Journal of Systems Science* published in London.

Just what has served to unite all these varied individuals and interests under the general systems banner is perhaps best expressed by George Klir:

> . . . we may say that general systems theory in the broadest sense has generated innovations in the following manner:
> 1. A new way of looking at the world has evolved in which individual phenomena are viewed as interrelated rather than isolated, and complexity has become a subject of interest.
> 2. Certain concepts, principles, and methods have been shown not to depend on the specific nature of the phenomenon involved. These can be applied, without any modification, in quite diverse areas of science, engineering, humanities, and the arts, thus introducing links between classical disciplines and allowing the concepts, ideas, principles, models, and methods developed in different disciplines to be shared.
> 3. New possibilities (principles, paradigms, methods) for special disciplines have been discovered by making investigations on the general level.[4]

In the simplest perspective (which, incidentally, we shall show is deceptively simple), the general systems theorist may be thought to be one who postulates isomorphisms among fundamentally different phenomena and, therefore, searches for analogy-based principles or models to explain behavior of 'classes' of entities or phenomena. Another aspect of interest is the fact that there is probably no scientist in this field who did not begin as a specialist in some specific discipline, only subsequently moving to adopt the general systems theory perspective. In this sense, interdisciplinary potential is both real and realized.

Yet the foundations being laid by the current interdisciplinarians may serve to support a class of professional systems theorists in the near future. For, as Zadeh and Polak have pointed out:

. . . system theory is a discipline in its own right—a discipline which aims at providing a common abstract basis and unified conceptual framework for studying the behavior of various types and forms of systems. Within this framework, then, system theory may be viewed as a collection of general methods as well as special techniques and algorithms for dealing with problems in system analysis, synthesis, identification, optimization and other areas within its domain.[5]

Now, it is the purpose of this book to explain, as fully as possible, the role which general systems theory can play in the social and behavioral sciences. And here we want to begin by avoiding the distinction between social and behavioral phenomena, preferring to treat the several disciplines of these separate sectors under one generic rubric: the *human sciences*. This allows us to combine the fundamental work of psychologists and psysiologists on the single individual, with the traditional sociological, political, economic, and anthropological emphasis on humans in collectivities. This simply reflects the empirically-predicated model which recognizes differences between the behavior of single individuals and groups, true, but also recognizes that there is a relation between singular and plural behaviors.

At any rate, we shall be working through a wide range of topics, all of them related to our purpose. Initially, there is the necessity to develop as clearly and precisely as possible the epistemological implications of general systems theory. In so

11

doing (hopefully), the reader will be alert to what is effectively unique in the general systems theory platform: its reliance on a holistic perspective; its concern with isomorphisms and macrodeterminacy; its employment of taxonomic and ideal-type constructs of both theoretical and pragmatic significance; the concept of the Janus-faced model or allegory, with one face turned toward the empirical justifications for the construct and the other toward the higher-level conceptual frameworks with which it must articulate; its demand that 'truth' be searched for in the interstices between successively more specific deductions and successively more general inductions; its maintenance that the 'organic' ideal-type is the proper referent for the human sciences, and that 'organisms' are unique in that they are never simple sums of their parts. To stress these innovative points of departure (at least as an epistemological 'set') is our mission for the first chapter.

Once we have the parameters (and promises) of general systems theory firmly in mind from our first chapter, we can move on to a fairly detailed defense of the ontological significances of the general systems theory platform for the human sciences. Now this is not as desultory as it sounds, for the problems associated with the loci of knowledge, with what is real and what is delusory, are very real for even the most casual scientist—and these points have been especially ill-developed within the human sciences. Specifically, by way of stressing the importance of this rather theoretical subject, we think we can show that the premises of many grand theories or important paradigms of the human sciences owe their origin not so much to any factual predicates, but to axiological predicates derived from empirically transparent assumptions about the world and everything in it: what Stephen Pepper has called world-hypotheses and what we shall refer to as *Weltanschauungen*. We ignore these obscure and often intangible predicates of our theories and concepts only at the peril of our scientific effectiveness and rectitude—and we would ignore them only against a specific dictate of the general systems theory approach which demands (as does common sense) that the premises on which our theories, constructs, and models are built be fully explicated and made opaque. There is also another aspect to this. Listen to Gerald Weinberg:

> . . . students find it hard to understand why Newton's calculations of planetary orbits is ranked as one of the highest achievements of the human mind. . . . But the general system theorist understands. He understands because it is his chosen task to understand the simplifying assumptions of a science—those assumptions which delimit its field of application and magnify its power of prediction.[6]

Thus, our journey into the fundamental ontological and epistemological bases of the general sciences, in our second chapter, will hopefully serve a dual purpose: To make clear the importance of explicating fundamental premises in *any* scientific endeavor that pretends to approximate reality (or intends to manipulate it); and the indication of how critical the ability to manipulate abstracts is to the ultimate success of any scientific enterprise. Hopefully, also, it will make clear that both the physical and human sciences have a long history, and that if we see at all clearly today, it is only because "we stand on the shoulders of giants."

In the third part of our work we want to take a look at the parameters of the human sciences themselves: just what are the factors, if any, which make our subjects different from those dealt with by the physical and natural sciences? What are the magnitudes of these differences? Do they legitimate the rather astonishing differences in methodology found between the physical and human sectors of modern science? What are some of the more predominant 'scientistic' traits we can identify within the human sciences? How can general systems theory, as a procedural platform, act to counter these symptoms of scientism? What effect will these counterings have on the contributions of the human sciences to society at large? What kind of a role does conceptualization play in the social sciences? How can we bridge the gap between theory and practice and, in the process, help heal the breech between the grand theory-builder and the empirical hypothesis-tester—the uncommunicative poles of the human sciences? In short, here we are concerned about the *relevance* of general systems theory.

In the fourth and final chapter, we shift our perspective. Here we shall be concerned about the specific role which the general systems theory approach can play in the social and behavioral sciences on the operational dimension—about the efficiency of general systems theory. Specifically, we shall develop a taxonomy of phenomenal ideal-types each of which is then associated with a specific subset of instruments drawn from the analytical arsenal of the modern sciences—and each of which is especially susceptible to one or another of the four major analytical modalities of science: (a) Positivism, (b) Inductivism, (c) Deductivism, (d) Heuristicism.

More specifically yet, we will be trying to open the methodological parameters of the human sciences to empirical examination, something Dewey would have applauded and something which the general systems theorist cannot avoid doing if his impact is to be positive. Beginning with some fundamental propositions about information economics, and employing the concept of the 'learning-curve,' we can begin to get a grip on some fundamental issues, especially those concerned with the limits of knowledge we can legitimately expect to obtain in the human sciences. In short, methodological issues may be at least partially lifted from the realm of rhetoric to which they have so long been condemned.

In a more general sense, then, we shall be trying to develop the general systems theory 'concept' to the point where it may be recognized by the practicing human scientist as a specific instrument of analysis on one level, and on a broader level, as the 'skeleton of science' postulated by Kenneth Boulding, which:

> . . . aims to provide a framework or structure of systems on which to hang the flesh and blood of particular disciplines and particular subject matters in an orderly and coherent corpus of knowledge. It is also, however, something of a skeleton in a cupboard—the cupboard in this case being the unwillingness of science to admit the very low level of its successes in systemization, and its tendency to shut the door on problems and subject matters which do not fit easily into simple mechanical schemes.[7]

As a final note, we sadly suggest that there are occasions when general systems theorists get carried away and, in the process, adopt an evangelical approach to their field which they so condemn in others. In truth, the ontological, epistemological, and procedural properties of general systems theory must remain working hypotheses whose utility and rectitude must be established by individual scientists within the confines of their particular interests. Clearly, however, general systems theory (as with any metaparadigm) would at least ask that the scientist's experimental and operational confines reflect the precepts set out. But so far as our work here is concerned, we will only attempt to provide whatever empirical or logical justification we have been able to isolate for the general systems theory cause within the human sciences, culled both from our own experiences in the field and in the reported (and referenced) work of others. In short, we shall try to defend the hypothesis that general systems theory offers an alternative to existing or traditional analytical platforms and that, as an alternative, it promises greater effectiveness, efficiency, relevance, and potential for substantive insight. But there can be no defense quite as plausible as validation under fire in the various sectors of the society asking assistance from the disciplines we represent.

An exposure to the evangelical extremes of general systems theory, then, would lead to an unfortunate reading of its ambitions and/or assertions. The casual reader may be led to the belief that general systems theory is capable of making General Motors intelligible in the same terms as an amoeba colony, or human behavior comprehensible in terms of classical mechanics. But this is not the aim of any reputable systems theorist. Rather, the general systems theorist is constantly honing Occam's Razor, searching for causal engines which transcend specific instances and yet, in application, prove more efficient as analytical 'masks' for the investigator to wear than those successive generalizations of the inductivist-empiricist (who the strict phenomenologist also declines to give much credit). Thus, where experience proves appropriate or to the limits that emerge as feasible, the general systems theorist wears an a priori 'mask' that concentrates on the possibility of morphological-dynamic commonalities among entities traditionally deemed 'unique' from other perspectives. But he is aware of this mask, and his presumption about its validity rises or falls to the extent that the empirical investigations he undertakes support or derogate it. His approach to a problem is dictated primarily by the emergent properties of the phenomenon itself, not (hopefully) by any a prioristic or axiological predicates which serve to artificially constrain the reality at hand or which predetermine the results of analysis. And this responsiveness to reality, this lack of theoretical or operational constraint—and the humility which these necessarily entail—are perhaps the highest recommendations which can be offered for the general systems theory platform.

1

General Systems Theory:
Parameters and Promises

Introduction

In the broadest sense, General Systems Theory is a supradiscipline, including such *special* system disciplines as mathematical system theory, systems engineering, cybernetics, control theory, automata theory. In another sense entirely, general systems theory offers a vocabulary of both terms and concepts applicable to systems of all types, with the terms and concepts drawn from many different substantive disciplines (i.e., biology, engineering, economics, quantum physics). In this chapter, we shall be concerned with yet a third and far more fundamental role for general systems theory: that of an essentially new epistemology of special significance to the social and behavioral sciences.

As an epistemological platform, general systems theory offers a set of precepts which, as an integrated set, is effectively unique. Among these precepts we would include the following:

· A belief in the scientific utility of (and the limited existence of) phenomenal isomorphisms.
· A belief in the prerequisital nature of proper theory, coupled with the conviction that empirical validation should be the arbiter of scientific truth.
· A postulation of the critical role to be played by analogic models in complex phenomenal domains.
· A preference for 'organic' referents within the social and behavioral sciences (as opposed to the mechanical referents of classical physics, etc.).
· A preference for a holistic analytical modality (predicated on hypothetico-deductive operations) as opposed to the reductionist-inductivist modality.
· A demand that instances of macrodeterminacy among complex 'organic' phenomena be fully exploited.
· The postulation that ideal-type and taxonomic constructs are the

most efficient vehicles for phenomenal analysis in the social and behavioral sciences.

When we speak of a general systems theorist, then, we are speaking first of all of a man who practices a specific substantive discipline (e.g., psychology, economics) but secondly of the disciplinarian who approaches his subjects using this epistemological platform, whose utility we will now explore.

1. A Strategic Appreciation

Almost twenty years ago, the program of the Society for General Systems Research was established, allocating to itself the following functions:

- To investigate the isomorphy of concepts, laws, and models in various fields, and to help in useful transfers from one field to another.
- To encourage the development of adequate theoretical models in fields which lack them.
- To minimize the duplication of theoretical effort in different fields.
- To promote the unity of science through improving communication among specialists.

The key here is the search for isomorphisms among real-world phenomena for these, when identified, permit the development of explanatory models or allegories via analogy-building. This allows us to replace several parochialized models with a single higher-order model, thus contributing directly to the *efficiency* of the disciplines involved. This is especially important for those disciplines whose subjects are inherently highly complex, for the analogy seems to provide the type of vehicle necessary for de-empiricalization, a hallmark of the mature science and the prerequisite for interdisciplinary interchange.
A. Isomorphy and Analogy

The general systems theorist's concern with isomorphisms must not be misunderstood. He does not postulate a Plotinian world, where all phenomena are deterministic and ruled according to a set of invariant, timeless, causal algorithms. But neither can he appreciate the now popular phenomenologistic view which finds the world and everything in it to be unordered, nonteleological and unintegrated, such that the dominant property of any empirical phenomenon is its overwhelming uniqueness. In the former world hypothesis, which underlies the Newtonian-Laplacian concept of phenomena as special instances of some mechanical ideal-type, the quest for universals via isomorphisms is so intense and compelling that science runs the danger of becoming little more than abject procrusteanism. In the phenomenologistic view, science is largely irrelevant, for the possibility of a concatenative, integrated body of knowledge is deemed a priori gratuitous; rather, knowledge becomes a personalized matter, with reality existing only in the idiosyncratic nexus between observer and localized object.

To a certain extent, both of these extremes are present in the social and human sciences, and have contributed largely to the problems which now afflict our disciplines. Particularly, the failure to find an acceptable middle ground for analytical and procedural activities, the failure to mediate between the grand theories of the Plotinian perspective and the trivializing, atomistic exercises of the positivistic counterparts of the phenomenologists, has lent the social and behavioral sciences all the characteristics of an abjectly immature science: (a) We have large numbers of localized, effectively uncorrelated (i.e., insular), and context-dependent empirical studies, largely unresponsive to any particular theoretical envelope; (b) This is coupled with the somewhat embarrassing existence of competitive, mutually-exclusive theories, each purporting to be *the* explanation for a single phenomenon; (c) These tend to become evangelistic rather than scientific plat-

21

forms, lending intradisciplinary communication something of the character of rhetorical argumentation rather than objective disputation.

Given these properties, it is perhaps too optimistic to suggest, as has James Conant, that the social and behavioral sciences are roughly in the same shape today as were the biological and medical sciences of a century or a century and a half ago.[1] What is really more to the point is the question as to whether we are really making any effort to get out of what Kuhn has called the immature, *pre-paradigmatic* days of a science,[2] or whether we are simply going to perpetuate the schism between deductively-oriented grand theory and artificially 'closed' hypothesis-testing which now finds our disciplines effectively polarized. In short, we lack the 'theories of the middle range' which Merton so consistently demands and which seem to have been absolutely fundamental to the success and progress of the natural and physical sciences. Lacking these, Merton echoes our schismatic interpretation of the human sciences, here with respect to sociology:

On the one hand, we observe those sociologists who seek above all to generalize, to find their way as rapidly as possible to the formulation of sociological laws. Tending to assess the significance of sociological work in terms of scope rather than the demonstrability of generalizations, they eschew the "triviality" of detailed, small-scale observation and seek the grandeur of global summaries. At the other extreme stands a hardy band who do not hunt too closely the implications of their research but who remain confident and assured that what they report is so. To be sure, their reports of facts are verifiable and often verified, but they are somewhat at a loss to relate these facts to one another or even to explain why these, rather than other, observations have been made. For the first group the identifying motto would at times seem to be: "We do not know whether what we say is true, but it is at least significant." And for the radical empiricist the motto may read: "This is demonstrably so, but we cannot indicate its significance."[3]

From this analysis it is fairly clear what characteristics a theoretical construct will have to have in order to act as an effective link between these two methodological poles. First, the construct will have to be Janus-faced, one face looking for points of articulation with higher-order constructs (i.e., meta-theories) while the other face looks constantly toward those derivable sub-hypotheses which will serve to permit empirical validation or invalidation. Secondly, the components of the construct should be connected according to laws of formal logic rather than elliptically or parabolically (as has been so often the case with the theories of the human sciences). Thirdly, the properties assigned the components of the model should be, as fully as possible, cast into effectively eidetic or universalistic terms, leaving idiosyncratic or subjective interpretations for the poets and rhetoricians.

Now when the analogic models which isomorphisms permit are constructed according to the criteria just set out, their promise in the social and behavioral sciences is immediately apparent. But the relationship between the identification of isomorphic properties among phenomena and the employment of analogies is not always the direct one that we expect. For example, there is a general tendency to assume that structural isomorphisms imply analogous causal factors. This is an assumption borrowed from the engineering sciences and classical physics, and one which is very dangerous within the human sciences. For example, both biological and social organizations tend to be isomorphic in this respect: both appear to accommodate increased complexity by evolving a hierarchical structure (or, inversely, the adoption of a hierarchical structure promotes complexification). Yet such structural isomorphy may not be anything except a delusionary prima facie case for causal isomorphy, for both biological and social phenomena share a critical property: the capability for exhibiting *equifinality* (the ability to arrive at the same structural state via different causal trajectories). When this capability is indeed exercised by a system, the presumption that structural isomorphy implies causal analogy is potentially misleading. As Anatol Rapoport has pointed out:

Insights derived from speculations instigated by perceived analogies function somewhat like education: they reveal to the intelligent and conceal from the stupid the extent of their own ignorance. . . . Therefore, seek simplicity and distrust it.[4]

This is excellent advice for those who would employ analogies, for though the mechanistic world hypothesis once had application among the essentially simple phenomena postulated by classical physics, its introduction into the social and behavioral sciences has had an obscurant effect (often resulting in the artificial-analytical simplification of real-world entities which can only meet the mechanistic criteria when constrained by entirely fabricative assumptions; as with the behaviorist attempt to legislate man as the sociocultural counterpart of the engineer's finite-state machine, or the popular conception of the business organization as the reification of the precepts of microeconomic theory or the equivalent of a digital computer).

Nevertheless, the search for analogies promises to be one of the most fruitful exercises which the social and behavioral scientists can undertake, for the analogy, if properly appreciated, may very well be the most efficient route toward the encompassing paradigms which our now atomized disciplines so desperately require. So, aside from the caveat about the a priori association of structural isomorphisms with analogous causality, the general systems theorist may be properly categorized as a scientist interested in isolating isomorphisms and following such isolations with legitimate analogies as initial working-hypotheses.

Thus, in stating the role for mathematical general systems theory within the social and behavioral sciences, Rapoport simultaneously states the case for properly-conceived analogic models in general:

> I submit we must face the fact that there are no biological or social 'laws' that are direct analogues of the laws of motion, the law of gravity, the conversion laws of energy and mass, the law of increase of entropy in isolated systems, etc. At most, there are *models* of specific biological or social phenomena, expressible as mathematical formulae to serve as working hypotheses. The models bring some simplification into the study of phenomena to which such models are reasonable approximations. If it happens that several phenomena of widely different content can be described by some mathematical model, both simplicity and generalization have been effected. Therein lies the power of *mathematical* general systems theory, based on the principles of mathematical isomorphisms, the completely rigorous version of analogy.[5]

Even granting the problems of casting social and human phenomena into terms which permit rudimentary mathematical treatment, the concept of the mathematical analogy is extremely appealing. It will still have allegorical implications, such that its utility will be found in its ability to accurately reconstruct historical system states or predict future system states. Now, clearly, where the same fundamental model may do this for two or more empiricals, and produce results which are neither so vague as to be otiose or so specific as to be open to charges of procrusteanism, then we have not only made a start toward *simplification* (in that we have de facto created a phenomenological ideal-type where before there were simply unrelated empiricals), but we have also introduced an element of *efficiency* into the scientific discipline involved. For, in the most basic terms, efficiency in a science simply means being able to explain the widest range of phenomena with the fewest unique models or allegories. As Ernst Mach has suggested: "The purpose of science is economy of thought."

It needs hardly be mentioned, as a point for closing this section, that there are good analogies and bad ones. As an

25

example, the now popular effort to use the cybernetic process or servomechanism as an initial heuristic in constructing *normative* models of enterprise behavior is an undoubtedly good use of analogy. Being normative, it is also a relatively safe use, well in keeping with the role of analogic engines as working-hypotheses rather than platforms for dogmatic inferences or in-context scientific description. On the other hand, an example of the ill-considered use of analogy is the behaviorist tendency to *describe* or explain human behavior in terms of the 'mechanistic' ideal-type derived from classical physics via Newton and Laplace.

In this case, the analogic device becomes transmogrified from a heuristic into an a priori predicate . . . an array of first premises which are insulated from empirical validation but which nevertheless are used to direct research and evolve prescriptions. As appealing as the mechanism might be as an analytical artifice (and as appropriate as it surely is to phenomena as physics used to define them), its employment as an engine for analogies within the social and behavioral sciences has been historically counterproductive, and continues to be so. For, as Karl Deutsch has rightly pointed out, the mechanical ideal-type carries with it some logical conclusions which are effectively inapplicable to our current knowledge of social and human phenomena, among them:

. . . the notion of a whole which was completely equal to the sum of its parts; which could be run in reverse; and which would behave in exactly identical fashion no matter how often these parts were disassembled and put together again, and irrespective of the sequence in which the disassembling or reassembling would take place. It implied consequently that the parts were never significantly modified by each other, nor by their own past, and that each part once placed in its appropriate position with its appropriate momentum, would stay exactly there and continue to fulfill its completely and uniquely determined function.[6]

The implications of the mechanistic ideal type are such, then, that its employment within the human sciences would be as much a symptom of procrusteanism as is the zoomorphism which attempts to make human behavior solely intelligible in terms of animal behavior or the anthropomorphism which sees the natural world in terms of humanistic counterparts and ignores or rationalizes away empiricals which cannot be so explained. At the extreme, then, analogy-building can take on procrustean characteristics, just as the a priori denial of any structural isomorphisms or causal analogies leads us into the scientifically sterile world of the phenomenologist. It is in the interval between the extremes, however, that the general systems theorist elects to work. And indeed, it is there he must work, for his ambition is to introduce that 'economy of thought' which is the chief property of the epistemologically healthy scientific discipline.

B. *Toward Theoretical Adequacy*

This 'economy of thought' can really only be realized by the process of *de-empiricalizing* a scientific discipline in its applied aspects. This requires some explanation, for it is a key proposition (or promise) of general systems theory. Initially, one of the most highly respected historians of science, James B. Conant, reproduced a quotation from an address given in the late 1800s by John Tyndall:

Hitherto the art and practice of the brewer have resembled those of the physician, both being founded on empirical observation. By this is meant the observation of facts apart from the principles which explain them, and which give the mind an intelligent mastery over them. The brewer learned from long experience the conditions, not the reasons, of success.[7]

Conant's interest in this statement is that it has a direct implication so far as the efficiency of a science is concerned, and he makes the case that the theoretical paradigms or models supplied by Pasteur and his successors greatly lowered the degree of empiricism in the manufacture of beer and wine. In short, the brewer or vintner no longer had to wake up in a new world each morning, where each unprecedented perturbation demanded new cycles of costly, time-consuming and uncertain trial-and-error. In short, it is no credit to a science that each of its subjects must be treated as a unique phenomenon, that all its parameters and states must be established empirically.

In all fairness, it must be admitted that the bacteria involved in fermentation processes are probably better behaved than the human beings involved in social phenomenon— better behaved in the sense that the bacteria have what appears to be only one possible reaction to any given action (or stimuli). But it is also clear that before Pasteur provided the theoretical envelope which generalized the conditions of fermentation, the centuries of empirical trial and error were not sufficient to explain the range of events which this single construct encompassed.

No stronger indictment of the excessive empiricism proposed by Bacon (and echoed by Hume) need be made. True, in the early days of the industrial revolution, the scientist was largely content to simply collect and catalogue data or to identify crude correlations which served some local purpose (i.e., the empirically generated 'If-Then' modality by which the reason for the correlations was considered less important than the fact of the correlation). In a delightful book, Frederick Vivian chides Bacon in the following way:

Bacon assumed that it was possible to use induction as a mechanical process. All we needed to do, he thought, was to observe and let the facts speak for themselves; observation would automatically lead, by the process of induction, to true generalizations. But facts do not speak for themselves; they must be arranged by a speculative mind. . . . In order to reach an understanding of the facts, we must be able to relate them to one another and see how they fit in with what we already know. To succeed in this, we need the gift of being able to make an imaginative jump from the facts we observe to a general theory, or hypothesis, which will explain them. . . . All great scientists have in a certain sense been great artists; the man with no imagination may collect facts but he cannot make great discoveries.[8]

By way of illustration, Vivian notes that biology before Darwin consisted mainly of unintegrated, unordered facts. "What Darwin did," he tells us, "was to change the emphasis in biology from induction to deduction; he moved from the simple search for more facts to the formulation of a general theory or hypothesis which could serve as a basis for further deductions."[9] We may also call on another science historian, Errol Harris, for a further opinion on the *primacy of the concept* in general science:

That scientists do not work in an intellectual vacuum or produce theories from the recesses of their minds *ab initio* . . . we . . . might well have expected. But neither do they extract their theories simply by empirical generalization from laborious collections of detail. No science is possible, no research can be conducted and no advance can be made except by reference to, and subject to the requirements of, some conceptual scheme.[10]

In a reference to Darwin again, Harris maintains that Darwin's theory could not have been inferred or produced as a direct mapping from empirical data. It was, he maintains, the relationships among the various data which were critical, relationships which could not have been observed without the at least partially formed hypothesis already held as a priorism by Darwin.

A fundamental advantage of a theoretical construct such as that developed by Darwin, leaving aside for the moment the question of 'how' he arrived at it, is in the nature of the sub-hypotheses which may be generated and which, in turn, provide foci for empirical validation of the envelopic concept itself. With the deductibility of these empirically susceptible sub-hypotheses, the purpose of grand theory is vindicated in the ways we expressed earlier. First, the theory may serve as a paradigm in Kuhn's sense, under which many different scientists may work, especially those who lack the conceptual equipment of Darwin himself.[11] Moreover, the paradigm provides the foci for points deserving further elucidation or

29

validation when there is a formal (i.e., nonelliptical) connection between the subhypotheses and the theory as a whole. In short, Darwin's great scheme served to structure and discipline biology; and even if portions of the theory itself have been derogated by further analysis and empirical evidence, there can be no question of its *efficiency* as a scientific vehicle.

Now we, here, are less interested in the grand conceptual schema of the natural or physical sciences than in whatever human science counterparts we might identify. For example, Keynesian economic theory, Marxian socioeconomics, Weberian axiology, and Freudian and behaviorist psychology have all sought to play a role similar to that of Darwin's construct. And all have provided at least semblances of a proper paradigm, lending some a priori coherence to otherwise ill-structured, diffused sets of phenomena. To this extent, the grand theories of the social and behavioral sciences have also served the role of 'de-empiricalization.' And they do this in essentially one (or both) of two ways: First, they provide a capability for deducing some 'events' which would have been empirically transparent in their absence (and which could not have been converged on by trial-and-error methods). Secondly, they helped focus the work of the empirical scientists away from simple description of isolated phenomena and towards formal analysis.

This, in turn, leads to another advantage of properly constructed theory: the linking of conceptual and experiential (i.e., empirical) efforts. Or perhaps we should say that, ideally, the properties of grand theory should be susceptible to experience and, on the other hand, empirical efforts should be responsive to theoretical direction. Such Janus-faced constructs then become the 'theories of the middle range' that men like Merton and van Nieuwenhuijze have so strongly advocated.[12]

The fact is, however, that whereas most major paradigms of the physical and natural sciences can be shown to have advanced the cause of 'economy of thought' and served as cohering forces, the paradigms of the human sciences' disciplines have often had an opposite effect—that is, they have sometimes served as platforms for intradisciplinary division. Rather than unifying a discipline in the pursuit of the validity or invalidity of a particular theory, the existence of one seems to spawn the existence of an essential inverse. For example, the psychologies of Skinner and Maslow are predicated on virtually antonymical premises, and the premises of both theoretical systems are equally empirically unvalidated. As such, they tend to constitute rhetorical rather than scientific systems, per se, and tend to be defended evangelically rather than nomothetically.

Similar disparities appear in the model bases of anthropology (i.e., evolutionistic vs. functionalistic predications), economics (fiscal vs. market-oriented theories of economic development), sociology (mechanistic vs. organic appreciations), etc. It is not, then, that the social and behavioral sciences lack theory, per se, but rather that we lack theory of the proper kind. If we follow the Marxian lead, our theories tend to be predicated on unvalidated first premises borrowed out-of-hand from some predecessorial philosophy or metaphysics; if we follow the Skinnerian lead, our theories tend to be predicated on some empirically unvalidated ideal-type such as his concept of 'non-autonomous man'; Maslow and Weber, on the other hand, offer us a model-building reference that smacks of subjectivism, such that the fundamental terms employed and the functional connections among their state-variables are often elliptical or downright parabolic in nature.[13] As such, large numbers of theories in the social and behavioral sciences tend to become little more than idiographic pretentions, evangelized by the schools' apostles on rhetorical grounds, attacked by members of opposing 'schools' on recriminatory grounds.

As to the type of theory proper to an emerging discipline,

and the type of theory which we have explicitly associated with the general systems theory platform, we can again listen to James B. Conant:

All the sciences concerned with human beings that range from the abstractions of economics through sociology to anthropology and psychology are, in part, efforts to lower the degree of empiricism in certain areas; in part they are efforts to organize and systematize empirical procedures. Whether or not in each of the divisions or subdivisions a Pasteur has yet arisen is not for me to say. But if he has, his contribution has been the introduction of some new broad concepts, some working hypotheses on a grand scale that have been fruitful of further investigations. It would seem important to distinguish, if possible, the advances connected with such broad working hypotheses, which are the essence of a science. . . . Many social scientists, I imagine, would not dissent too strongly from the proposition that their whole area of investigation is in a state comparable to that of the biological sciences (including medicine) a hundred or a hundred and fifty years ago. If this be the case, the balance of this century should witness great strides forward; but by the very nature of science (as compared to empiricism), it is impossible to foretell in what precise direction the advance will be made.[14]

But while we were waiting for our Pasteurs, the general systems theorist can help fill the interval most profitably. He will not, however, make his contribution by developing another psychology to stand beside those of Maslow and Skinner; rather, he will be working at a new level of the scientific hierarchy. Particularly relevant to the social and behavioral sciences is the fact that the fundamental perspective of the general systems theorist would all but forbid him building anything except interdisciplinary constructs—unless a specific real-world phenomenon could be shown to be entirely determined by variables drawn from a single academic discipline. And when these interdisciplinary constructs are properly conceived and executed they would provide a framework within which conflicts among lower-level theories could be resolved. Indeed, logic will tell us that when we have two 'peer' theories, each purporting to treat a specific phenomenon in competitive ways, there will be some higher-order construct which can be imposed on the adversaries such that the variables of the lower-order theories will be subsumed by the variables of the higher-order. In other words, for example, there will be some point of abstraction (and this point need not be at a meaningless level) where the divergent trajectories of Skinner and Maslow may be seen to share a common node. And, for those familiar with their respective theses, this node would be found somewhere in the work now being done by psychophysicists and learning-theorists or developmental psychologists such as Piaget.[15]

Simply, within the operational domain of the human sciences, the general systems theory potential is to be looked for in its ability to assist us in developing higher-order conceptual envelopes which serve to encompass (and supersede) the competitive, lower-order theories of scholastic advocates. The construction of such models is, then, the subtler of the two ways in which the general systems theorist attempts to achieve the third objective of his society: minimization of the duplication of theoretical effort in different fields. The second and more obvious way is through the mechanism of analogy-building when significant points of isomorphy are identified among sets of phenomena.

In all, then, the four major strategic precepts of general systems theory may be rephrased as follows: (1) The dictate that a complex phenomenon should be initially approached with the intention of identifying whatever points of isomorphy may exist between it and other better-known phenomena (on either the structural or functional dimension); (2) The employment of properly constructed (and constrained) analogic models following the recognition of points of isomorphy—this is a fundamentally different approach than the use of an analogic model as a priorism; (3) A belief, historically supported and documented, in the prerequisital nature of theory and

33

deductive constructs employed as initial heuristics; (4) a dictate that theoretical constructs should be Janus-faced in the sense that we have mentioned, and that components should be entered and connected nonelliptically and nonparabolically. We must now become more explicit.

2. The Array of Operational Prescriptions

In addition to the strategic guidelines mentioned above, the general systems theorist brings to his discipline or areas of investigation the following operational tenets:

- The postulate that nonmechanical 'wholes' are not simple sums (or products) of the properties of their parts.
- A preference for a holistic as opposed to a reductionist analytical modality when treating the 'organic' or 'open' systems which predominate in the social and behavioral sciences.
- The concentration on macrodeterminacy as an isomorphic property of many (or most) complex systems, and as a fundamental point for analytical departure.
- The employment of ideal-types, taxonomies and typologies as the fundamental vehicles for the advancement of science in complex phenomenal domains.

We shall now briefly explore each of the operational tenets in some detail, hopefully indicating their points of articulation with the strategic precepts we earlier set out.

A. Wholes and Parts

As to the first dictate, let us listen to Ludwig von Bertalanffy:[16]

The Aristotelian dictum of the whole being more than its parts, which was neglected by the mechanistic conception, on the one hand, and which led to a vitalistic demonology, on the other, has a simple and even trivial answer—trivial, that is, in principle, but posing innumerable problems in its elaboration:

The properties and modes of action of higher levels are not explicable by the summation of the properties and modes of action of their components *taken in isolation*. If, however, we know the *ensemble* of the components and the *relations existing between them*, then the higher levels are derivable from the components.

We here have the postulation that different levels of any hierarchy of complex character tend to be governed by factors (e.g., laws, algorithms, principles) which may themselves be different. Therefore, on the behavioral or functional dimension of a system, we must expect qualitatively new factors to be introduced as we move to new levels of ordering—factors which cannot, except in the most unusual circumstances, be *induced* from the behavioral properties of the system's several parts. In this sense, then, wholes are granted a phenomenal 'personality' as it were; and this personality is unlikely to be a product of aggregation or simple multiplication.[17]

But this does not imply that *some* systems we shall encounter in the domain of the human sciences will not be effective aggregates. Indeed, as the anthropologist Leslie White has pointed out, there are instances of social entities which have evolved via the process of *segmentation*, such that each component of the system is a structural and functional replicate of every other.[18] Here we would have the human sciences' counterpart to the amoeba colony of the ecologist, though these acephalic systems will be only rarely encountered. When we *do* find a system which has tended to emphasize segmentation as opposed to differentiation, however, we can be a priori more certain that the following conditions will hold:

· The entity known as the 'whole' will bear a strong morphological/functional relationship to the individual parts in isolation.
· Unobserved portions of the entity will tend to be morphological/functional replicates of the empirically observed.

In such a case, properties associated with the whole will tend to be aggregative (such that, for example, input-output parameters will vary directly with the number of components encompassed by the whole at a given point in time, etc.).

In general, because most social and human systems tend to emphasize differentiation as opposed to segmentation (as, indeed, do most biological systems of any sophistication), we will usually have to be aware of the potential for each new level of an entity evolving properties unique to it. Or, at the very least, we would have to approach any phenomenon except an essential mechanism as one which inheres this capability or potential, and therefore be extremely cautious in using the inductive modality subsequent to reduction of the entity.

Granting wholes a potential for hierarchically-oriented uniqueness (following from structural and functional differentiation), we come to a critical analytical branching point. For, if wholes are not simple aggregates or products of the properties of their parts, the information gleaned about the whole from analyses of the parts would be highly equivocal at best, downright erroneous at worst. In short, we would expect to incur a significant *error in synthesis*, where synthesis simply involves the upward integration or assemblage of lower-order parts and properties. This brings us squarely against the reductionistic modality.

B. Reductionism and Reductability

A question here must immediately come to mind: what if we were able to isolate and study *all* components of a system in a reductionist framework? Would not, under these very favorable circumstances, the error in synthesis be reduced to a tolerable level? We must answer this question in two ways, for it is critical to our work here. On the one hand, we again must face the problem of concatenative differentiation as a human sciences' phenomena—or the problem of complexification in general. Specifically, if we had an empirical knowledge of *all* the properties of *all* the system parts, and a knowledge of *all* possible relationships among them, would we have a knowledge-set which is identical with a knowledge of the entity as a whole? We quite obviously would not—at least not unless all system action stemmed solely from internal forces. What we would still need for a comprehension of the entity as a whole is some kind of *field-function*, a knowledge-set which expresses the relationship between the exogenous factors present in the system's environment and the properties of the system's parts. Thus, to explain the system, we must go outside the system.[19] This, in essence, is the concept behind the effort to have human scientists adopt the view that the predominance of our subjects are *open systems*, entities whose viability depends on a continuous interchange with successively more encompassing systems. Only the deliberately insulated, artificially fabricated system of the finite-state engineer can be understood as a totality by an analysis of its own constituents. For the human scientist, the problem appears to be one of being able to adapt our methodology and our perspective to appreciate the necessity for expanding our research space. As Kurt Lewin suggested many years ago:

Observation of social behavior is usually of little value if it doesn't include an adequate description of the character of the social atmosphere or the *larger unit of activity* within which the specific social act occurs.[20]

What all this means is simply this: the tendency to try to take social systems apart, or to partition human behavior into isolated compartments, is scientistic. For, except for the very rare entity which will meet the criteria for the essential mechanism, behavioral properties of any real significance are generally dependent upon a context for their viability. From the system perspective, this means that the search for the whole cannot end simply with an inventory of parts analyzed in effective isolation or with partitioned expressions of their interrelationships. Rather, the *field* must also be defined. And it is the system as a whole, not the isolated parts, which is significant when we move to the ecological dimension of sociocultural or behavioral phenomena. As such, the complex system tends to become a *temporal gestalt*, carrying on interchange with other temporal gestalts within a field which, itself, must be defined and treated as a gestalt rather than a mechanism.

Aside from this problem of the ecological field not being adequately treated by examinations of system parts, there is also a very practical problem: for anything except the most simple system, it will be virtually impossible to accurately allegorize or model the range of interactions and interfaces available to system components. As such, the question about whether or not a knowledge of a whole can be built up from a knowledge of *all* properties of *all* parts in *all* potential relations becomes gratuitous in the domain of the human sciences. In this respect we can listen once again to Ervin Laszlo:

> Of course, it is quite possible that we could fully account for the properties of each whole if we could know the characteristics of all the parts and know in addition all existing relationships among them. Then we could reduce the characteristics of the whole to the sum of the characteristics of the parts in interaction. But this involves integrating the data not merely for three bodies, but for three thousand, three million, three billion, or more, depending on the whole we are considering. And since science cannot perform this feat even for a set of three parts, it is quite hopeless to think it can do it for any of the more complex phenomena it comes across in nature, man, and society. Hence, to all practical purposes, the characteristics of complex wholes remain irreducible to the characteristics of the parts.[21]

Here, then, we add a practical argument (to stand beside the theoretical one proposed earlier) against the effort to induce properties of wholes from properties of parts without incurring a significant error in synthesis. To these we may add a third argument, a substantive one.

There is the problem that many (if not most) phenomena we will encounter in the human science domain will behave organismically—where the properties assumed by any part are both nondeterministic and nonexclusive. That is, many systems are so effectively integrated that there is simply no 'part' which can be abstracted from the whole without losing significance. This 'organic' ideal-type, as a scientific reference, is as old as Epicurus (for whom body and soul were both corporeal and coterminus so that all reactions of an individual were psychosomatic).[22] For Radcliffe-Brown, a system meant 'a complex organized whole,' whose parts were so interrelated that one would alter their natures if one attempted to abstract them from the system of which they were parts.[23] Clearly, this organismic perspective lends the social psychoogy of Lewin its impacts as it does that of Maslow and the gestaltists. And not surprisingly, one of the strongest advocations of the organismic perspective comes from the great biologist, René Dubos:

In the most common and probably the most important phenomena of life, the constituent parts are so interdependent that they lose their character, their meaning, and indeed their very existence, when dissected from the functioning whole. In order to deal with problems of organized complexity, it is therefore essential to investigate situations in which several interrelated systems function in an integrated manner.[24]

Thus, to some of the most eminent theorists in widely different fields, the phenomena of the human sciences become instances of 'organized complexity' where the method of organization is organismic. Though we shall have more to say about this later, the organismic ideal-type suggests a reflexivity among system components, a reflexivity so intense that the part removed from context for analysis becomes a fiction of no ontological (i.e., real) significance.

C. The Holistic Potential

In its fundamental aspects, the holistic modality applauded by the general systems theorist does not deny either the value of empirical analysis or the occasional reduction of entities for scientific manageability. It, simply, demands that *some* awareness of the whole precede the attempt to appreciate the parts. In other words, some comprehensibility of the total character of a complex system is a prerequisite for meaningful and efficient research in the human sciences. Without this initial 'mask' we can wear in approaching complex and ill-structured phenomena, without some concept of the field and of the organizational structure which integrates the parts, we emerge with vast amounts of uncorrelated data and isolated studies whose applicability in an 'organic' world is sorely tried.[25] In short, we emerge with the atomization we earlier associated with much of the human sciences.

Most fundamentally, the general systems theorist tries to approach complex phenomena with a prior (and necessarily crude and tentative) comprehension of the entity as a whole, a comprehension which will serve the role of a flexible, empirically-responsive heuristic.

This initial heuristic then serves as a reference for the selection of variables for empirical analysis, with the proviso that the results of the empirical analysis be fed back to modify the original heuristic. The critical difference between this approach and more traditional methodological platforms predicated on reductionism (i.e., that of classical physics), is that a disciplined and constrained *hypothetico-deductive modality* displaces the inductive.[26] The constraint or limiting factor of this approach, then, lies not so much with the complexity of the subject being studied, as with the imagination or breadth of knowledge possessed by the observer (or by the team of observers). The observer's prior experiences, his ability to manipulate abstracts, his sensitivity to gestalten which may be present in an a priori unorganized field, his ability to develop meaningful analogies—all these move to complement the traditional empiricist skills of attention to detail, scientific skepticism, and nomothetic precision.

In one sense, then, the holistic platform is an attempt to reunite the conceptual skills of the philosopher with the mechanical and instrumental skills of the laboratory scientist, a union which has been ill-obtained since the empiricist platform came to dominate science (while Bacon dominated epistemology and the industrial revolution dominated society). For the technics of classical physics simply do not maintain themselves well beyond a certain level of phenomenal complexity. The strategy of varying one factor at a time, the concept or artifice that 'all other factors can be held equal,' the process of concatenative induction from one system level to the next—these have served mechanics and electrochemistry rather well in their applications but have had an insidious and disintegrative effect on the human sciences.

Rather, as Walter Buckley points out: ". . . the modern systems approach aims to replace the older analytic, atomic, Laplacian technique with a more holistic orientation to the problem of complex organizations."[27] Going somewhat further, Buckley cites the disfavor which currently attends traditional analysis (where one factor at a time is varied, under the *ceteris paribus* assumption), introducing a quote from W. Ross Ashby suggesting that the way *not* to approach a complex system:

. . . is by analysis, for this process gives us only vast numbers of separate parts or items of information, the results of whose interactions no one can predict. If we take such a system to pieces, we find we cannot reassemble it.

In another place, Ashby is more specific:

. . . in the 1930's, general system theory arose, mostly through the work of Ludwig von Bertalanffy, who saw not only that the study of parts (in "classic" science) must be supplemented by the study of wholes, but also that there *exists* a science of wholes, with its own laws, methods, logic and mathematics. Also working in the 1930's, R. A. Fisher had appreciated how fundamentally limited was the approach through analysis, taking one variable at a time . . . studying all the variables, one at a time, would exhaust many lifetimes, and . . . the method of examining the parts individually

41

was fundamentally incapable of giving information about the interactions between the variables. . . . Thus arose one of the basic contributions to the modern epistemology of "the system." It was based, it should be noted, not on the *results* of field experiments but on getting clear, *before an experiment was started*, what could give information about what.[28]

Here, then, is a strong voice for the heuristically-driven holism that promises to accomplish tasks in the complex regions of the human sciences that reductionism (and the consequent closed-system analysis techniques) either chose to ignore or simply failed to bring to successful conclusions.

D. *Macrodeterminacy*

The other factor which makes the holistic approach particularly promising is the phenomenon of *macrodeterminacy*. This is a situation which occurs when a system may be treated as deterministic at the higher levels (or at the level of the whole itself) but where the lower-order components of the system may not admit to determinacy. In the analytical domain, this may mean that the system itself is capable of prediction (or amenable to some sort of finite-state system analysis technique) even though we cannot treat the parts or lower-order components as deterministic. The implications of macrodeterminacy for the human sciences is easily as great as for the physical sciences. In fact, macrodeterminacy may be one of the fundamental isomorphisms which the general system theorist may be able to isolate among all complex 'organic' systems. Its importance for us, however, is perhaps best explained by Paul Weiss, a biologist of great moment:

If physics has had the sense of realism to divorce itself from microdeterminism on the molecular level (with the displacement of classical physics by quantum physics), there seems to be no reason why the life sciences, faced with the fundamental similitude between the arguments for the renunciation of molecular microdeterminacy in both thermodynamics and systems dynamics, should not follow suit and adopt "macrodeterminacy" regardless of whether or not the behaviour of a system as a whole is reducible to a stereotyped performance by a fixed array of pre-programmed micro-robots. Since experience has positively shown such unequivocal macrorelation to exist on various supra-molecular levels of the hierarchy of living processes in the documented absence of componential microdeterminacy, we evidently must let positive scientific insights prevail over sheer conjectures and pre-conceptions, however cherished and ingrained in our traditional thinking they may be.[29]

The implications to be drawn from the phenomenon of macrodeterminacy turn directly toward the concept of the holistic methodological approach just outlined. Particularly, to the extent that macrodeterminacy exists, holism is not only the most promising approach but the most potentially efficient as well. It acts on the supposition that one need not be able to allegorize the behavior of parts in order to determine (or predict) the 'state' of some system as a whole—in instances where this is indeed a valid presumption, the holistic modality of the general systems theorist becomes expedient as well as instrumental.

It is important to appreciate just how fundamental macro-determinacy is to the human sciences. As was earlier suggested, when systems under treatment begin to depart from the most simple kind of mechanisms, both our mathematics and our logic (even coupled with a large computer) fail to be able to manage the interactions and componential permutations of behavior potentially available. And inasmuch as virtually no significant (i.e., nontrivial) human science subjects meet the criteria for the essential mechanisms of the classical physicists,[30] macrodeterminacy becomes the *sine qua non* of economics, social psychology, political science, anthropology, etc.—as these are all disciplines which owe their existence to the possibilities of making (reasonably) valid and precise assertions about entities which are too complex to be analyzed in detail or made internally deterministic. Clearly, if it were not for the tacit existence (or presumption) of macrodeterminacy among the subjects of the human sciences, the only discipline which could exist with any legitimacy at all would be that type of psychology which, under the phenomenological epistemology, would reserve itself to making point-in-time statements about single individuals, ignoring collective or ramifying behaviors entirely.

E. The Taxonomic Potential

Given the concept of macrodeterminacy (and its immanence within the domain of the social and behavioral sciences), the construction of ideal-type and taxonomic models appears to be the most promising methodological trajectory we can follow. The preference for these is a reflection of the general systems theorist's concern with isomorphisms and the holistic analytical modality, as they are an especially good way of approaching a priori ill-structured, highly complex phenomena.

Essentially, the construction of proper taxonomies and ideal-type models is an exercise in the hypothetico-deductive process we mentioned earlier. An *ideal-type*, initially, is an abstraction which emphasizes certain properties which are felt to be distributed among real-world entities, and emphasizes them in such a way as to produce a highly directive and normative model. The model is directive (if not operationally, at least experimentally) in that it provides the major premises from which some logical conclusions may be drawn; these logical conclusions suggest what we would expect to find in the real world *if* our ideal-type has any empirical validity. So long as the logical conclusions (and we may call them subhypotheses) are drawn with logical validity, and so long as they are phrased in such a way as to permit empirical validation via field experiments, then we have the connection between abstract theory and empirical experience that we would seek in a conceptual structure.

A properly formulated ideal-type will be intentionally abstract and encompassing in nature. That is, it will try to establish an analytic category which will hopefully encompass an array of empiricals, such that a large number of phenomena may be viewed (holistically) as special cases of some generalized system. We thus combine the search for isomorphisms, the demand for a holistic perspective and the postulated efficiency of a priori theory in the proferring of the ideal-type as a scientific vehicle. The ideal-type should also permit the derivation, under the rules of formal deduction, of subhypotheses which are susceptible to empirical validation, and susceptible in such a way that the results of empirical trials will reflect in some calculable way (e.g., statistical) on the probable validity of the ideal-type itself. In this latter sense, the ideal-type and its subhypotheses perform the role which Braithwaite set out for deduction within science: the ideal-type itself becomes the set of 'initial propositions' and the subhypotheses, when formed according to the transformational rules of deduction, become conclusions.[31] Moreover, the entire system becomes reflexive, such that validation on one level tends to impact on another. As such, proper theory, derived from the platform of the ideal-type, *itself* takes on the characteristics of an organically interconnected, hierarchically organized system.

The relationship between an ideal-type, properly formulated, and a taxonomy is simply this: an ideal-type is one element in an array of ideal-types which, as a set, constitutes a proper taxonomy. This is not to say that some ideal-types cannot exist independent of a proper taxonomy (e.g., Weber's 'bureaucracy'); yet, clearly, the directivity or insight lent by an ideal-type is strengthened if it is one among several *intervals* or points on a continuum. Even a dichotomy is analytically preferable to the isolated ideal-type. In every sense, then, the kind of taxonomic construct the general systems theorist would prefer is that which is constructed around a single set of dimensions (properties), with broadly different qualitative or quantitative values on those dimensions yielding the different ideal-types. When this is accomplished, the subhypotheses associated with a taxonomy will reflect (in terms of their formally deduced predictions) degrees of difference which reflect the differences in positioning on the continuum occupied by the several ideal-types—such that, for example, the subhypotheses associated with the two ideal-types at the poles of the continuum should produce predictions or prescriptions which are themselves polaric, etc.

While the human scientist without much interest in epistemological problems may find proper ideal-types and proper taxonomies (as we have defined them here) difficult to identify within his own discipline, he will have less difficulty if we suggest that most of the *typologies* of the social and behavioral sciences are really ideal-type constructs—with the dimensions of the typological construct capable of being transformed onto some sort of continuum. But the typology, as it has long been practiced in the human sciences, provides an additional capability: it may serve as a structure for uniting the product of two continua, or even three or four continua, such that we emerge with matrices of multivariate importance. The cells of a properly constructed typology may serve as shared points between several different taxonomies, such that we perform a qualitative 'mapping' of complex fields in much the same way that a topologist may work with the constructs of abstract algebra.

Indeed, if we take the liberty of extending the work of Anatol Rapoport, we find him heading toward exactly this position:

The task of general system theory can be formulated as follows: to prepare definitions and hence classifications of systems that are likely to generate fruitful theories. . . . There is a connection between this task and the one . . . of counteracting the increasing factionation of science.[32]

One may extend this remark to suggest that these classifications may serve this purpose of integrating the human sciences especially when they take the form of typologies *which serve to unite taxonomic constructs from different disciplines.* Ideally, the field of socioeconomics, social psychology or other deliberately interdisciplinary areas might well look to this type of typology as the central vehicle in their development, with the cells of the resultant matrices being at the interstice between continua from sociology and economics, etc.

Beyond these specific considerations, Rapoport serves to distinguish between a good classification and a poor one:

Good classifications are those that are likely to produce concepts from which a far reaching theory (in the strict sense of the word) can be constructed. A poor classification is one that has no such far-reaching implications.[33]

We, as human scientists, do not have to look too far for instances where vast and imposing classification schemes fell short for just this reason. An excellent example is Maurice Mandelbaum's criticism of the theoretical 'systems' developed by both Malinowski and Radcliffe-Brown: ". . . the attempt to establish laws of social organization on the assumptions of either Malinowski or Radcliffe-Brown ended in failure: Their generalizations permitted no deductive consequences with respect to the specific nature of the practices of the peoples which they, and other descriptive anthropologists, investigated."[34]

We may summarize much of this by simply suggesting that the general systems theorist, when he enters any disciplinary domain, will be primarily interested in examining existing theoretical constructs—and in developing new ones—which fully reflect the criteria for the hypothetico-deductive methodology—and that, in most cases, the ill-structure and inherent complexity of human science subjects will demand initial attack via ideal-type, taxonomy, and eventually the interdisciplinary matrices which take the form of the typological 'map.'

3. The Interdisciplinary Potential

None of the elements associated with the general systems theory platform is, of itself, unique and unprecedented. The holistic approach is very much in the spirit of the intuitionalistic preference of the German Romantics, eventuating in the grand theories of Marx and Weber. It is distinct from what has been called the 'discursive consciousness,' the hallmark of empiricism, in that the latter demands that we proceed from specifics to generalities—the reification of the inductive method. As Karl Pribham reports:

In the case of discursive consciousness of a whole or of a complex of any sort, definite consciousness of the whole is subsequent to the consciousness of the parts, whereas the opposite procedure obtains when intuitive consciousness is applied. . . . Hence, all those who endorse the intuitional pattern have this much in common: they emphasize the power of the human mind to grasp the essential characteristics of complex phenomena in their totality and to segregate these characteristics from accidental or fortuitous features in order to arrive at "insight" into the true nature of events.[35]

Yet, were we to review the history of the intuitive consciousness in practice in the history of the human sciences, we find that these 'insightful' models seldom were phrased in ways which permitted empirical validation in the spirit of Popper's "open Society," and seldom were they strict adherents to the logical discipline inherent in the hypothetico-deductive methods recommended by the general systems theorist.

Similarly, taxonomical and typological constructs have abounded in the social and behavioral sciences; in fact, it is not too much to suggest that the majority of the information we think we possess in these fields is tied up in one or another of these ideal-type derivatives. But, again, if the logic is loose and the ideal-types are forgotten for what they really are—

hypothetical premises—they tend to become little more than footnotes to social history, bereft of any lasting scientific potential. They serve rather more as rallying points for evangelists than as the grand paradigms which have stood the physical and natural sciences in such good stead.

The same thing stands for the quest for isomorphisms—often they have tended to produce models which are blatantly procrustean, as with the Hegelian-Marxian attempt to report all empiricals as derivatives of some strict dialectical process or with the Spencerian attempt to explain all social behavior in terms of biological referents. They are interesting and useful, true, but their utility is always dimmed by the fact that the subhypotheses derived from such structures tend to be insulated from the original premises, such that empirical validation or invalidation has little impact on the theoretical predicates themselves.

From the epistemological perspective, then, the general systems theorist is fundamentally more disciplined and cautious than the idiographic-rhetorical grand-theory builder, and fundamentally more imaginative and open than the tester of localized, nonramifying hypotheses. On the ontological dimension, he will search for potential sources of knowledge in two domains: (a) The domain of empiricals as the positivists and phenomenologists would recommend; (b) In the realm of concepts, as the intuitionalists who inherited the Cartesian and Kantian mantles would suggest. But, aside from such generalisms, we can be explicitly concerned about the role of the general systems theorist in promoting interdisciplinary interchange or in achieving the fourth and final parameter of the Society For General Systems Research program: increasing communication among specialists from widely diverse fields.

We have already seen how the general systems theorist seeks to instill 'economy of thought' within a discipline; it should also be obvious how he seeks the same objective between or *among* disciplines. Basically, interdisciplinary interchange will depend upon the extent to which isomorphisms may be found among the subjects of the respective academic areas (and upon their depth). This, as we have already seen, will determine to a large extent the legitimacy of replacing parochial (i.e., intradisciplinary) models with generalized, analogically-driven ones. Now, the recognition of isomorphisms and the generation (and/or identification) of analogic models of appropriate resolution in their turn depend largely upon the general systems theorist's ability to arrive at a *substantive vocabulary* that will do for the social and behavioral sciences much of what the *abstract vocabulary* of mathematics has been able to do for the natural and physical sciences.

Thus we are beginning to find conceptual links being forged and operational hypotheses being exchanged among members of many different human science disciplines: sociology, economics, social psychology, anthropology, political science, management science, organization theory, etc. Buckley, for example, as a sociologist, uses systems terms such as *feedback, open systems, entropy, equilibrium* and *steady-state*, etc., terms whose implications are immediately apparent to economists such as Boulding, to political scientists like Easton, to behavioral scientists such as Berrien or Trist, to psychologists such as Rapoport or to management scientists such as Churchman, Ackoff, or Simon. In the absence of this substantive vocabulary which general systems theory provides (along with the various special system theories such as cybernetics, mathematical systems theory, automata theory, etc.), the extent of functional interchange among these disciplinarians would surely be more limited than it is today—and the social and behavioral science disciplines would still be treating their subjects as if they were actually partitioned along the same lines as our college faculties.

The general systems theory potential, at least on this dimension, is certainly not restricted to the social and behavioral sciences however. The attraction for the natural or physical scientist is somewhat different though: the implication-filled terms of general systems theory can act to complement the interchange which mathematics has already fostered, and to move communication toward a new level of specificity and operationality. The other dividend from the general systems theory vocabulary (of terms *and* concepts) derives directly from this kind of interest: it permits intercommunication between the social-behavioral and natural-physical sciences as the broadly-defined sectors of modern scientific enterprise. For the physicist and biologist and chemist are also becoming increasingly familiar with the substantive as well as the mathematical implications of key elements in the general systems theory vocabulary. This has a powerful benefit for all concerned, for this common array of terms now permits the search for fundamental system isomorphisms to extend beyond sectoral rather than just disciplinary bounds. For, clearly, the physical and social sectors have been too long separated; and this separation has been largely at the expense of the social and human subjects which we, as human scientists, must strive to serve, not simply understand.

At any rate, we have now exhausted the broader aspects of general systems theory as a fundamentally new epistemological platform for the social and behavioral sciences. It now is necessary to explore some of the more basic ontological implications of this new approach for, if for no other reason, epistemology should always be subordinate to a formal ontology of some sort. In the chapter which follows, then, we shall be paying explicit attention to what general systems theory has to say about scientific rationality, and about knowledge in its most basic sense.

2

The Rationalizing Role of General Systems Theory: Its Ontological Implications

Introduction

We have seen that general systems theory constitutes an effectively unique epistemological-methodological platform for the social and behavioral sciences. But, to this extent, it must also have some very fundamental ontological implications which underlie its procedural significances. These were introduced somewhat indirectly and casually in the first chapter, for it was there that we mentioned the importance of the Janus-faced theoretical construct—one face searching for points of articulation with higher-order conceptual structures; the other face looking explicitly for ways in which the speculative or deductive hypotheses may be empirically validated or invalidated.

Such constructs represent something of a departure from traditional theoretical or conceptual devices. In the following statement, Ludwig von Bertalanffy explains why this is so and, at the same time, sets the ontological stage for us:

> In one way or the other, and with whatever minor modifications, *all* theories took for granted the Cartesian dualism of matter and mind, things and consciousness, object and subjects, *res extensa* and *res cogitans*; accepting them as indubitably given and trying to bring them into some intelligible relationship. By now, however, it has become obvious that neither "matter" nor "mind" stood up to the test of scientific investigation. . . . Analysis has to proceed at two levels: that of *phenomenology*, that is of direct experience, encompassing perception of outside things, feeling, thinking, willing, etc.; and of *conceptual constructs*, the reconstruction of direct experience in systems of symbols, culminating in science; it being well understood that there is no absolute gap between percept and concept, but that the two levels intergrade and interact.[1]

It is the implications of this remarkable statement which we want to explore in this second chapter of our work, especially as its tenets affect the performance of the human sciences.

1. The Array of Alternatives

Epistemological and ontological issues are no longer the province of the intellectual dilettante or the abstruse philosopher. Scientists have come to realize that they set, in many cases, the premises under which investigation, analysis, and model-building take place—in effect, they are what we might loosely refer to as *transparent axiological predicates* of scientific enterprise. This can hardly fail to be the case, when we consider the fundamental and imminently important issues which are entailed in the epistemological-ontological arena. Just how fundamental they are is indicated by this statement from William Gondin (written, incidentally, over thirty years ago):

Perhaps the most prevalent conception of the central problem of epistemology is still largely dominated by such questions as whether a mind's cognitive states can relate to anything beyond the immediacy of its own consciousness, or whether sense data are driectly relevant to the nature of Reality, etc.[2]

Such questions are of epistemological significance largely because, depending on the answer we give, our fundamental analytical approaches to science will be different. For example, if we look at the two polar positions on the continuum of epistemological platforms, *universalism* would answer yes to the first question, no to the second, while the *nominalists* have historically answered just the other way round.[3] It is equally important to note the primacy of the ontological issues entailed here, for the questions cited above have explicit implications as to the locus of Reality—where we should begin to look for truth, as it were. And the saddest fact of all is that these questions have not been answered in any satisfactory way in the intervening thirty years since Gondin wrote (much less in the two thousand or more intervening years since such questions led to the effective partitioning of Greek philoso-phy).[4]

Moreover, it is difficult to see just how we are going to be able to yield satisfactory answers to these questions any time in the near future. There are, however, some points of indirect evidence. For example, it is difficult to explain the origin of mathematics except as an instance of hypostatization (i. e., reification of some order intrinsic to the mind); a similar claim for a priori ordering of some kind may be made with respect to Chomsky's postulation of immanent 'deep structures' and whatever evidence is beginning to amass regarding the ability to genetically transfer information.[5] Such points tend to make us want to give a tentative *yes* answer to the first question: 'whether a mind's cognitive states can relate to anything beyond the immediacy of its own consciousness.' In a somewhat similar way, phenomenologistic experiments aimed at measuring the isomorphic qualities of the senses (i.e., by asking different individuals for explicit descriptions of various empiricals) would tend to point toward a *yes* answer for the second question: 'are sense data directly relevant to the nature of Reality?' But modern psychology has given enough experimental evidence as to the idiosyncratic potential of the senses to suggest a *no* answer. In short, these fundamental ontological-epistemological questions, the ultimate predicates of science and knowledge, remain answerable only incompletely, ambiguously, and often simply through the medium of rhetoric. And, indeed, as I shall try to show, this is the way they have been answered historically.

From the crudest perspective, the universalistic thinkers (mainly theologistic scholars with their epistemology predicated on exegesis), have tended to consider sense data as delusionary—and, as we suggested, modern psychology of perception would give them some limited support in this. To a lesser extent, the Neoplatonists, Descartes, and Kant also tended to minimize the ontological significance of the sense datum, whereas the Germanic romanticists (of the Verstehen school) and Hegelians, etc., following Kant's lead, tended to make empirical reality either subordinate to or determined by conceptual-cognitive innates (i.e., the dialectic of Hegel was worn as an a priori *mask* which imposed a preordering and preselection on real-world phenomena). One must be somewhat careful here, for while all the positions tended to deny any ontological significance to the senses, their degree of optimism about the mind's ability to capture Reality varied greatly.

On the other historical extreme we have those positions which deny innate ideas, cognitive constructs or apriorisms any role in the pursuit of knowledge (or *scientia*, per se). The origin of such positions rests, or course, with the nominalists, but the cause was formalized and popularized in the prologomena to inquiry set out by Francis Bacon. Bacon, for example, preferred "that reason elicited from facts," and had a hearty and thorough distrust of reason *cum* metaphysics as a scientific vehicle (a distrust somewhat warranted by the excesses of the theologistic metaphysicians in preceding ages). Thus, Bacon accomplished the shift of science to inductive inference and Locke brought forth from this the ontological precept of the *tabula rasa*, with the direct corollary that all knowledge has its origins in experience.

In the Baconistic extreme, science should be hypothesis-free, leaving accumulated facts (i.e., sense data) to speak for themselves. In the more moderate form, reason has an a posteriori role to play—and only that role. It is used to engineer generalizations via inductive inference and hence give rise to "that reason elicited from facts." From the Baconian position, then, we can arrive almost directly at the hallmarks of modern science: (a) A belief that ontological significance dwells only with those phenomena which are entirely empirically accessible; (b) That empirically accessible phenomena may be effectively reduced for analysis in isolation and synthesized through successive inductive inference; (c) That real knowledge is only that which may be phrased in strictly eidetic, strictly nomothetic terms, and has no a priori existence in the form of prepercepts or preconcepts.

There is a third position we can isolate, one which is of some importance but of limited scientific relevance: the *phenomenologistic* platform. The origin of modern phenomenology rests with the extreme Baconist position just cited, mediated by Hume. It differs primarily in that, in addition to having a real abhorrence of a prioristic or deductive devices, it also questions (in its strict interpretation) the relevance of any attempt to treat phenomena as more than locally significant. That is, both universalism and generalizations via inductive inference are to be avoided, and Reality rests somewhere in the nexus between observing subjects and context-bound objects. There is, of course, a similarity between this position and that of the proper existentialist, for whom knowledge was always a personalized matter, incapable of being transformed out of its subjective aura, and for whom attempts to share knowledge (i. e., generalize or universalize it) were deemed futile. Hence, the existentialist philosopher and the phenomenologistic scientist share a penchant for idiographic, elliptical and/or parabolic constructs.

With these very brief comments, we have the basis for developing an array of epistemological/ontological ideal-types, as follows:

· The *Rationalist* platform, which stresses the significance of the mind (i.e., reason, cognition) as a source of Reality, and which stresses the existence of certain context-independent or universalistic laws or principles governing the behavior of empiricals— and which stresses the belief that these laws or principles must be deduced rather than induced (or derived from an examination of the empiricals themselves). The position may be extended as follows: some apriorism in the form of a deductive or theoretical 'mask' is required before any meaningful order can be made apparent in the empirical domain.

59

- The *Empiricist-Positivist* platform, on the other hand, demands that science confine itself to the realm of empiricals (wherein all ontological significance is deemed to reside), lending reason (i.e., the mind) only a posteriori significance in the search for meaningful (as opposed to random) patterns which may emerge from the results of sets of empirical analyses and investigations. Scientific truth, then, is only that which is the product of inductive inference operating on an empirical data base, and the only apriorisms which are to be tolerated are those hypotheses which are themselves of empirical origin.
- The *Phenomenologistic* platform stresses the sensate boundaries of Reality and, in the extreme, proposes that valid knowledge is always a personal and context-dependent affair, born in the gestaltlike nexus between subject and object; a prioristic instances (i.e., innate ideas; hypotheses; axiological predicates) must be guarded against, as introducing an illegitimate 'coloration' into empiricals. Similarly, attempts at generalization are delusionary, as the predominant property of any empirical phenomenon is its fundamental 'uniqueness.'

In short, then, all these positions exercise a determinacy on scientific enterprise (or those scientists following one or another of the platforms) in that they restrict the legitimate domain of inquiry and, in the process, restrict the nature of the outputs which result from scientific analysis. The first or *rationalist* position, a priori relegates the sense datum and the empirical phenomena susceptible to sensate capture to secondary importance, or in the extreme may deny them any significance whatsoever. The second and third positions, on the other hand, a priori deny reason any ontological significance except, in the case of the empiricist-positivist platform, an a posteriori 'ordering' role.

Now the fundamental reason why the general systems theorist may be displeased with these positions is simply a reflection of a point made earlier: we simply do not know, with any degree of acceptable certainty, the relationship between reason and reality or, for that matter, between sense data and reality. As such, these platforms remain largely just products of speculation or rhetorical bias. Therefore, it is extremely dangerous (if not downright short-sighted) to confine science to one or another of these domains, to a priori restrict our pursuits of scientia when the restrictions and constraints (lacking empirical validation) must remain simply matters of axiological or personal preference. Moreover, and this is especially key here, we cannot prove or disprove the relationships between reason and reality or between sense data and reality by *purely logical* or axiomatic means. Neither Kant nor Bacon succeeded in this, nor have any of their modern predecessors. In summary, then, there is as good a reason to accept one of the platforms as another, and no reason at all to accept one at the direct expense of another.

In this sense (although, in all fairness, the epistemological/ontological positions remain fabricative straw-men), there is every reason to develop some sort of *triadic ontology* to supersede, as an operational platform, these several unsatisfactory dyadic ontologies.

Simply, the ontologies just outlined are dyadic in that they employ the Cartesian dualism of subject and object, percept and concept, cognition and empirical. And, because they all represent logically possible solutions to the ontologic problem, none can be out-of-hand rejected except when it becomes the *sole* engine of analysis. Therefore, the rationalizing influence of general systems theory is expressed in its adoption of a triadic ontology which subsumes the relevant portions of all three traditional positions. In its simplest phrasing, we can suggest that there is no significant evidence to a priori restrict reality to either the cognitive, empirical, or subjective (i.e., idiosyncratic) domain, and there is every evidence to suggest that knowledge (i.e., scientia) may be some product of all three domains. Therefore, the rational ontology would be one which considers the hypothesis that any quantum of

61

Truth may be due to some interaction of cognition, empirical observation and subjectivism—at least potentially. Simply, the rational ontology would be one that is entirely a priori unconstrained. And indeed, this kind of postulate seems entirely in sympathy with the position taken by von Bertalanffy with which we began our work here. The next task, then, is to examine each of these positions in more detail, expanding on some of the general criticisms we earlier made, and pointing out those functional or positive aspects of each platform which should be incorporated in the general systems theory triadic ontology.

2. The Rationalistic Position

A good example of the rationalistic extreme is Marx's attempt to a priori legislate socioeconomic behavior by the imposition of the Hegelian dialectic. Less extreme instances are the attempts of Herbert Spencer to make social behavior a direct reflection (i.e., special case) of organic or biosystemic analogies or the Freudian attempt to make human behavior dependent upon the stochastic flux of unconscious forces in the mind, drawing their energy from the libido, etc. The link between these efforts is their postulation, in the form of apriorism, of some causal principle which gradually acquires the nature of a 'mask' to be worn which lends an order to real-world phenomena which cannot (or have not been) encompassed empirically in their totality. The mask not only serves an ordering function, but tacitly restricts observations or pursuits of evidence to phenomenal properties which are congruent with the apriorism (and may act, in many cases, to filter out evidence incompatible with the theory or paradigm at hand).

An interesting property of most rationalistic constructs is this: they are not derivable, in all their aspects, from the empirical phenomena which they purport to explain or describe. They usually entail an additional component which might variously be referred to as intuition, imagination, mysticism, revelation, etc. To this extent, the constructs of Freud, Spencer, and Marx all bear the stamp of hypostatization which, depending on the individual's *feelings* toward the schema, may indicate either genius or chicanery, insight or delusion. These properties lend rationalistic constructs another important aspect: while they all are functional so far as explaining some of the properties of the empiricals they attempt to treat, none of them can explain *all* the properties. In short, in the logical sense, they are *insufficient*. In other words, only part of their predictions are valid and, as such, only parts of the theory can be correct.

Such constructs have a debilitating influence on the disciplines which permit them (along the functional influences which even a partially-correct theory entails), largely because they tend to cause the discipline to split into warring camps. On the one side are those axiologically (or emotionally) enamored of the theory, and whose major scientific efforts seem to be largely devoted to its evangelization; on the other hand will be those who use the existing theory primarily as a vehicle for generating an inverse theory, which may partially explain why we tend to have *two* psychologies and *two* anthropologies, etc. And then there is the fundamental danger associated with all rationalistic constructs: they tend to move science away from the validation or invalidation of the fundamental principles or predicates on which the theory is founded (and these, indeed, may be transparent) and turn it instead in the direction of making the generalizations more specific—that is, in deducing subhypotheses which then become the subjects of experimentation, subhypotheses whose connection with the first premises of the theory is either elliptical or indirect such that the grand paradigm itself is effectively insulated from experience. This, I feel, is also the problem that Thomas Kuhn associated with many grand constructs which he would consider proper paradigms, per se.[6]

But Kuhn, as well as such philosophers of science as Errol Harris and James Conant, are not shy about reciting the historical benefits which science has derived from constructs such as Freud or Darwin or Marx developed (though recognizing that there would be qualitative differences between these and the paradigms lent science by, say, Pasteur or Einstein). Indeed, the work of all three of these scholars suggests that the deductive construct in the form of grand theory or proper paradigm is a virtual *sine qua non* for scientific progress—or, more metaphorically, that if empiricism, reductionism, and inductive inference are the vehicles of science, the fundamental trajectory which they must follow is set by deductive inference and the speculative foundations which ultimately underlie it.[7]

Once again, however, the danger is that the premises tacitly driving the paradigm or theory may escape empirical attention. This, obviously, has a potential benefit as well as a liability to offer, for it economizes on the energy and effort of those scientists in the majority in any disciplines—those who are not bred for original theory-building or those not adept at manipulating abstractions. The positive promise of the rationalistic construct is then found on the dimension of *efficiency*, as Kuhn here suggests:

When the individual scientist can take a paradigm for granted, he need no longer, in his major works, attempt to build his field anew, starting from first principles and justifying the use of each concept introduced.[8]

Now, while this has proven very beneficial in the natural and physical sciences, where the paradigms tend to be cast in eidetic and nomothetic terms and constructed in such a way as to be immediately susceptible to empirical validation or invalidation, one can hardly be so sanguine about their impact in the social and behavioral (or human) sciences. For here, grand theory tends to assume some of the characteristics of metaphysics, such that first premises are not introduced as hypotheses but as axioms, and many supposed syllogisms turn out to be enthymemes.

Along this line, Frederick Vivian has given us an illuminating analysis of the problems which can stem from excessively rationalistic theories, here with respect to the discipline of history:

> The empirical historian makes use of the inductive process to discover various historical trends; he may even attempt to find a pattern, or historical laws which will enable him to predict future events; but these patterns or laws will arise out of his knowledge of what in fact happened. The metaphysical historian, on the other hand, starts with a grand design already in his head into which historical events have to be fitted. He makes the assumption that history must make sense, and the most obvious way in which this can be done is by explaining individual events as part of a wider purpose, or plan. . . . When events are viewed in this light, they cease to be merely things that happened because of certain determining causes, and become inevitable and necessary as part of a pre-ordained whole. An historical explanation becomes no longer a matter of explaining *how* things come to pass, but of showing *why* they *had* to come about as they did. . . . The different patterns which Hegel and Marx professed to find in history were not discovered from the events themselves, but invented and then imposed upon history from without.[9]

But again, if some of the greatest deflections from reality have been owed to rationalist grand-theory building, so have some of science's grandest moments. For example, Errol Harris gives grounds for attributing empirically-independent (i.e., Neoplatonic) origins to Newton's fundamental laws, having Descartes serve the role of a deductive mediator:

> Descartes asserts that, just as a body retains its shape and other properties, so must it retain its state of motion (speed and direc-

65

tion), unless some external agency acts upon it. This follows from his principle of conservation of motion, which he claimed to deduce from the immutability of God. . . . Thus the total quantity of motion which God included in the universe is conserved . . . and every change of motion that occurs is compensated by an equal and opposite motion (cf. Newton's third law). If this is indeed the *proximate* origin of Newton's laws (for the ideas can be traced back much further), clearer demonstration that, historically, they were not empirically derived is hardly needed.[10]

Despite these eventually functional results from 'unbridled Neoplatonic speculation,' the rationalist position in the sciences cannot be defended in its entirety. In short, it would be ultimately as unprofitable to build an epistemology on the Kantian or Cartesian ontologies as it would be to build one solely on the Baconian position. With respect to the former, the problem arises because both Descartes and Kant were led into positions which required them, for the logical consistency of their respective metaphysics, to virtually ignore the reality potential of sense data and the empiricals these pertain to. Van Steenberghen explains:

. . . Descartes' work was carried on under conditions that caused him to lead modern intellectualism in the direction of rationalism and idealism. First of all, his historical information was very limited. He was not acquainted with Aristotelian or Thomistic realism. . . . In the second place, Descartes was strongly impressed by the discoveries of science. Physics had furnished much new and precise data on the nature of bodies. But instead of using these data to correct the traditional doctrines about sense perception, Descartes fell into complete scepticism regarding sensation. He no longer sees in sensation an immediate contact of consciousness with the external world; he regards human consciousness as a closed consciousness, without any direct relation with the realities foreign to the self.[11]

As for Kant, Van Steenberghen has this to say:

The ambition of Kant was to end once and for all the "dogmatic slumber" of reason (which had too much confidence in its own value), and to institute a genuine critique of knowledge. This critique would try to determine the conditions under which scientific knowledge was possible, and thus fix the limits of human science. At first sight, Kantianism seems to be an effort to reconcile empiricism and idealism, a patient attempt at reconstruction intended to restore unity to the edifice of knowledge. But in reality the value and function which he attributes to the *datum* are so reduced that the whole weight of reconstruction must be borne almost exclusively by the *activity of the knowing subject*.[12]

But, still in this context, the adherents to the rationalistic platform, especially in its demand for holistic, encompassing treatment of phenomena as a *prerequisite* for meaningful empirical or positivist analysis, might well tell their critics what Kepler wrote about Tycho Brahe over 350 years ago:

Tycho possesses the best observations, and thus, as it were, the material for the building of a new edifice; he also has collaborators and everything one could wish for. He only lacks the architect who would put all this to use according to his own design. For although he has truly auspicious talents and real architectonic skill, yet the multitude of the phenomena and the fact that truth is deeply hidden in the particular details are obstacles to progress.[13]

Finally, we might want to temper our adulation of the architectonic role of the grand-theory builder with these thoughts from C. Wright Mills:

When we consider what a word stands for, we are dealing with its *semantic* aspects; when we consider it in relation to other words, we are dealing with its *syntactic* features. I introduce these shorthand terms because they provide an economical and precise way to make this point: Grand theory is drunk on syntax, blind to semantics. Its practitioners do not truly understand that when we define a word we are merely inviting others to use it as we would like it to be used; that the purpose of definition is to focus argument upon fact, and that the proper result of good definition is to transform argument over terms into disagreements about fact, and thus open arguments to further inquiry.[14]

In this context, Mills is particularly critical of the work of Talcott Parsons and around others who have a preference for defining and employing terms in idiographic and often largely subjective ways. He also takes a quiet but sharp glance at the prevalence of concept-laden terms such as *capitalism, middle class, bureaucracy, power-elite,* and *totalitarian democracy.* Particularly of concern, if we interpret his criticism correctly, is the fact that such terms are not properly units of analysis for sociology or social psychology or political science, but obfuscations, conceptual clusters, which tend to turn human science model-building into ". . . more often an arid game of Concepts than an effort to define systematically—which is to say, in a clear and orderly way—the problems at hand."[15] More specifically, to the extent that the terms employed in grand theory in the human sciences are analytical as opposed to synthetic (and analytical without the grace of relationally-independent definitions), the theories they underlie are subject to charges of gratuitousness and subjectivism and obscurancy. What Mills did not say, but what we cannot ignore, is the possibility that we have so many 'arid' constructs in the human sciences for the simple reason that so many of their developers fail to realize that taxonomies and ideal-types *are themselves theories,* not simple definitional or relational structures—and that, as theories as opposed to expeditious artifices, they must adhere to the eidetic-nomothetic criteria that we rightly demand of any construct pretending to either operational or analytical directivity.

3. The Empiricist/Positivist and Phenomenologistic Stances

If the empiricist stance stands fundamentally opposed to the rationalistic, the phenomenologistic stands fundamentally opposed to the general systems theory platform. In either case, the empiricist and phenomenologistic platforms have a hearty and thorough distrust of reason as a route to scientific reality (i.e., scientia) and, in their extremes, see any attempt to impose a cognitive order on phenomena (a priori or a posteriori) as ontologically delusory. The empiricist and phenomenologistic patterns differ in very fundamental ways, as we shall see, yet both have adopted the nominalistic precept that the realm of the empirical is the sole residence of reality and the only admissible target for knowledge. And both despise metaphysical efforts.

Summing up the positivistic/empiricist distaste for reason *cum* metaphysics is this statement by Professor Hogben:

> Science occupies its present position of power and prestige because it has learned the hard lesson that logic, which may be a good servant, is always a bad master.[16]

But the metaphysician, in his turn, may have a few unkind words to say about nominalism's a priori restriction of objects of ontological significance to the empirical domain, thereby negating the potential contribution of deductive engination to scientific discovery. One of our most eminent ontologists, James Feibleman, had this to say:

> "Carnap in rejecting metaphysics said that he would confine to that topic those propositions which claim to represent knowledge beyond experience but not those which attempt to arrange scientific knowledge in a system, since the latter belong to empirical science and not to philosophy. What empirical science is it, we may ask, 'whose object is to arrange the most general propositions in the various regions of science in a well-ordered system'?"[17]

Feibleman goes on to suggest that this is the task of the metaphysician in his role as ontologist. More generally, however, this is a role which must be assumed by deductive reasoning, either by the philosopher or, as it is currently being done, by general systems theorists. Without the deductive models providing interdisciplinary interfaces among the several empirical

69

(and parochialized) scientific disciplines, the scope of knowledge potentially available to a concatenative, generalizing science is capriciously restricted. In his words:

In seeking to escape from the unrestricted subjective dogmatism of the metaphysicians of the past, the logical positivists have ruled against metaphysics in general and have thus fallen into the error of promulgating an absolute negative dogmatism. For by getting rid of an infinite metaphysics, they have omitted also all considerations of the possibility of constructing a system as comprehensive as the limits set by the demands of logic, mathematics and the empirical sciences will permit, and thus have prevented themselves from being led forward to the discovery of a finite ontology.[18]

Nevertheless, there is more than a grain of promise for the sciences in Bacon's less extreme postulations about the probity of strict empiricism, expecially when he goes after the Mosarabic metaphysicians or others who took the stance that argumentation (i.e., reason and recrimination) is the end-all of science. For in the *New Organon* we read:

It cannot be that axioms established by argumentation can suffice for the discovery of new works, since the subtlety of nature is greater many times over than the subtlety of argument.

Hence, the cornerstone of the "great instauration" and the initial engination for positivism via nominalism. But considering the political and economic implications of the Reformation, and the first glimmerings of the social implications of the Age of Reason on whose threshold he stood, it is not to be wondered that one entirely unvalidatable ontological preference would be called in to sweep clean an equally empirically inaccessible set of a priorisms which preceded it.

In the most elementary sense, the transition from the universalistic perspective to the nominalistic was accompanied by a radical shift in inquiry. Men began to abandon the cosmologically-oriented question of "why" things worked and began asking naturalistic questions about "how" they worked.

This was entirely in keeping with the coming industrial revolution, and with its coming the death-knell for metaphysics was sounded, our preoccupation with the natural sciences began and the virtual equation of 'science' with physics became a legitimate premise for ages to come. And, as could not have failed to happen, an epistemologist/ontologist emerged riding Bacon's horse who suggested that scientific inquiry confine itself to those elements of interest to the emerging elite of the late seventeenth century: the middle-class bourgeois. This, of course, was John Locke, who attempted to make science ". . . an adjunct of a routine of industrious attention to business."[19]

But Bacon was not without his critics, and the deductive potential was not to be done in because of its mistaken identity with the mysterious speculations of long-dead mystics. The Germans kept the equation between reason and reality alive in the work of Kant, and Descartes' concept of 'innate ideas' was never fully debunked by Locke except to Locke's own satisfaction. For Descartes, if we may be permitted some liberties, man's pure intelligence was a reflection of God's intelligence (even if an imperfect approximation). Mathematics was the purest instance of the correlation of creatorial and human properties. Moreover, mathematics (i.e., a logico-mathematical capability) could be conceived of as man's only innate defense against the deception of nature, as it was a direct reflection of the fundamental engine of causality. Therefore, the only true knowledge is that which was amenable to axiomatic allegorization. Kant took a similar view in its most fundamental components. As Maurice Natanson has suggested:

Kant sought to illuminate the architecture of the mind through a kind of inventory of forms and categories. One central result of this mode of philosophizing is that a philosophy of mind becomes a philosophy of consciousness, not a psychology of brain activity, but an exploration of the purely formative features of all percep-

71

transcendental philosophy, a style of analysis which is developed in nineteenth and twentieth century neo-Kantianism and phenomenology."[20]

Hence we have the reliance (ad extremum) on disembodied intellectual processes as the sole source of 'scientia,' and the inherent distrust of empirically accessible phenomena has an a priori predication that is both powerful and immanent in the philosophical constructs of the Germanic philosophers and the staunch advocates of the ontological equation of reality with the exercise of reason alone. That which is finite and that which is 'natural' automatically becomes dubious—if not because of the imperfection of the correlation between perceptions and reality, because of the failure of inferences drawn from a specific interval of time to reflect teleological reality. In the broadest sense, then, Descartes and Kant might speak for those who distrust not only empirical methodology but the entire predication of science on inductive inference, which is the hearthstone of the naturalistic-positivistic scientific community which predominates today.

Yet the arguments of Hume *for* the empiricist modality (or, more precisely, *against* the rationalistic modality) are both powerful and appealing:

"I shall venture to affirm, as a general proposition which admits no exception, that the knowledge of [any cause-effect] relation is not, in any instance, attained by reasoning *a priori*; but arises entirely from experience, when we find that any particular objects are constantly cojoined with each other. Let an object be presented to a man of ever so strong natural reason and abilities; if that object be entirely new to him, he will not be able, by the most accurate examination of its sensible qualities, to discover any of its causes or effects."[21]

Now, as a philosophical argument, Hume's position can be criticized only metaphysically. Yet, nevertheless, it has become reified into many *substantive* entities which greatly affect the sciences, and particularly the social and behavioral sciences. For example, the a prioristic arguments of Hume and Locke, when merged, provide the explicit foundation for what might be called the 'naturalistic' school in linguistics and the 'behaviorist' school in psychology. Both, in their own ways, consider the behaviors of interest to have natural or empirical roots, such that both language and human activity (and, from there, social activity) are solely products of experience (individually or collectively gained, and transmitted either via genetic processes or through the social medium of institutionalization, etc.) In much the same way, psychological *environmentalism*, where behavior is entirely context-dependent and localized in origination or motivation, and *functionalism* in anthropology are also reifications of this epistemological precept. In short, it has had tangible, directive and prescriptive determinacy for the modern sciences, even though it remains itself a speculation or metahypothesis.

Thus, Chomsky's annoyance that empiricism should be accepted as the "null-hypothesis" of science has a legitimate basis, especially within his own field of linguistics:

In short, it seems to me that the question of the basis for acquisition of knowledge is an open, empirical question, to be settled by empirical investigation rather than by *a priori* argument or by pure conceptual analysis. Specifically, the empiricist assumptions have no special status among the many theories that might be proposed for the acquisition of language, or anything else.[22]

Indeed, when one begins to search for *empirical* evidence as to the ultimate (and exclusive) rectitude of the empiricist-positivist platform, one is hard put. Hempel, for one, finds the whole issue of empiricism tied inextricably to reductionism as a procedural phenomenon, and reductionism, in its turn, tied to versions of the basic ontological questions with which we began our work here—questions which have never been answered, such as:

Are mental states nothing else than brain states? Are social phenomena simply compounds of individual modes of behavior? Are living organisms no more than complex physiochemical systems?

Are the objects of our everyday experience nothing else than swarms of electrons and other sub-atomic particles?[23]

Indeed, much of the work done by general systems theorists, or by other leading edges of modern science, explicitly contradicts such postulates through experimental evidence of unimpeachable empirical quality.[24] As such, the general systems theorist is most apt to appreciate the alternative to this reductionist perspective, one described by Hempel as the "doctrine of emergence" which, he suggests:

. . . would have it that as we move from subatomic particles to atoms and molecules, to macroscopic objects, to living organisms, to individual human minds, and to social and cultural phenomena, we encounter at each stage various novel phenomena which are irreducible, which cannot be accounted for in terms of anything that is to be found on the preceding levels.[25]

If we avoid, however, the adverse implications of empiricism and positivism when carried to extremes, there are several points of considerable advantage we must note as associated with this platform: (a) Initially, there is the admirable dictate that theories which pretend to descriptive or predictive significance should be entirely susceptible to empirical validation or invalidation; (b) Secondly, there is the understandable and often laudable preference for inductive as opposed to deductive theories, especially as the latter are amenable to the imposition of an objectively (i.e., statistically) determined index of accuracy or probability of error; (c) Thirdly, there is the *efficiency* of the reductionist-inductive inference modality in dealing with those phenomena which are effectively simple or 'mechanical' in nature; and, finally (d) There is the empiricist-positivist preference for logically direct, eidetic, and nomothetic models, properties which are absolutely preferable to the elliptical, subjective, and idiographic attributes which so often accompany rationalist constructs.

Yet, once again, the fundamental argument against the empiricist-positivist position is that it has often gone well beyond the simple advocacy of the criteria mentioned above and, instead, chosen to take the necessarily hypothetical belief in the empirical as the sole source of ontological significance and transmogrify it into a dogma—thus closing the sciences' door to that reality which may exist in the other domains. And, on a more general level, we find some problems with the empiricist-positivist platform on the substantive dimension, among them: the inability to adequately deal with phenomena which are empirically transparent or only indirectly accessible; the inability to deal with counter-factuals; the failure to provide an adequate mechanism for the analytical reconstruction of phenomena of strictly historical significance in the absence of contemporary data bases.

In closing this section, we want to say a few brief words about phenomenology and its ontological significance. Ironically enough, it was Hume himself who set effective limits on the significance of the empiricist-positivist platform and, in the process, called into question the probity of the inductive inference modality. Here is a particularly revealing statement:

> In vain do you pretend to have learned the nature of bodies from past experience. Their secret nature, and consequently all their effects and influences may change in their sensible qualities. This happens sometimes, and with regard to some objects: Why may it not happen always and with regard to all objects? What logic, what process of argument secures you against this supposition?[26]

The uncomfortable conclusion of this statement is that neither logic nor exhaustive empirical investigations can help us translate inherently stochastic entities into deterministic ones.[27] More specifically, we have here the rationale for comdemning all inductive inferences (whether temporal, as Hume treats, or cross-sectional/spatial) to the realm of the problematic. Quite properly, then, Hume foresaw the reason why the 'laws' of physics, mechanics, and chemistry, etc., would become statistical propositions rather than theorems, per se. And perhaps he anticipated, inadvertently, precisely the kind of conceptual engine which would force the transition from classical to quantum physics more than two centuries later. But more importantly for our purposes, it limits the potential effectiveness of natural science instruments and strict empiricism and inductive inference in treating social-science subjects. For to the extent that one accepts the well-documented case for human subjects inhering a capability for 'strategic' or opportunistic behavior (i.e., self-initiated change), then positivism must be properly restricted to providing descriptions of such systems or treating historical behavior, for its predictive significance will be sorely constrained.[28]

As we earlier suggested, strict phenomenology would make two critical assumptions which differentiates it from the empiricist-positivist stance:

(a) A fundamental disbelief that inductive (or for that matter, deductive) inference can lead to any nonlocal, ramifying principles or generalizations—i.e., there are definite spatio-temporal boundaries of ontological significance such that each phenomenon should be approached as effectively unique.

(b) A disbelief in the eidetic quality of knowledge—i.e., the assumption that all knowledge is personalized and obtainable only within the shifting and gestaltlike nexus between the observer and observed at a particular point in time (or space).

Strict phenomenology, then, would demand that we approach phenomena unfettered by any apriorisms in the form of theories, axiological predicates or hypotheses and without any teleological (or predictive) ambitions. For, at least as interpreted by Husserl, phenomenology is "an a priori science of essences" but it is *nondeductive*.[29] Thus, the knowledge which is a product of phenomenologically-driven enterprises is neither distributable, concatenative, nor universal. As Walter Buckley has pointed out:

The phenomenologist simply moves the focus from sense data to mental configurational entities as unique, private, primitive givens of experience . . . and appeals to private intuition as the arbiter of knowledge.[30]

Here, then, knowledge is always a product of experience (or, more precisely, of sensate happenings between object and subject), with the additional proviso that sense data become personalistically transmogrified in the process.

Even Husserl admits that we must go back to Hume for the fundamental origins of what has come to be known as phenomenology, but in the modern context, Hume's position has become mediated by Kant. Van Steenberghen explains:

Misled by the phenomenalism of Hume, he [Kant] considered the data which presents itself to human consciousness to be "pure diverse" lacking all internal structure and intelligibility. This automatically rules out any conception of human consciousness which would *assimilate objects*. And since he refused to think of human consciousness as a *creative* consciousness, he imagines a consciousness *producing its objects by "informing" a datum which of itself would be formless*. . . . The Copernican Revolution was complete. It was no longer the real which informed consciousness, rather it was consciousness which informed the real.[31]

In its strict form, then, phenomenology is irrelevant to the pursuits of science, yet entirely relevant to the existentialist, the poet or the mystic for whom knowledge is first of all personal, second of all ephemeral.

In its milder and more popular form, phenomenology has come to be associated with the intuitionalistic or *Verstehen* schools, about which Karl Pribham has had this to say:

The method used by the intuitional school for the establishment of social laws consists in generalizing inner experience—in grasping the essence of complex social phenomena by a sort of penetration into their true meaning (*Einfühlung*). German advocates of this procedure have suggested that the 'nominalistic' method of comprehending (*Begreifen*) used by the natural sciences in establishing social laws should be contrasted with the method of understanding (*Verstehen*) available to the "sciences of the human mind and of society."[32]

And, somewhat elliptically, when we begin to get an appreciation of the *Verstehen* approach, we find ourselves right back in the rationalistic camp, but this time even farther into the reaches of subjectivism and idiographic model-building. And so the circle is complete.

4. The General Systems Theory Platform

Through the explication of the rationalist, empiricist-positivist and phenomenological platforms we have tried to point out the danger of turning an ontological apriorism into a dogma, a dogma which will determine the subjects elected for study, the nature of the analytical approach and the character of the results of scientific enterprise. In short, ontological issues are *not irrelevant* for the practicing scientist. We have, however, tried to show this *logically* in preceding sections. Yet there is also considerable historical (and hence empirical) evidence that scientific progress is not in fact made *solely* in the empirical or *solely* in the rationalistic or phenomenologistic domains.

James Conant, for example, sees the great progress of the natural sciences, since the time of Galileo, as the result of a *convergence* of metaphysics and empiricism:

As I now read the story of the advances that have been made in the natural sciences since the time of Galileo, what has happened is essentially this. An age-old process of inquiry by which artisans and skillful workers improved the handling of inanimate nature became gradually associated with the type of thinking up to then characteristic of mathematics. Another way of putting it is to say that two streams of human activity, separated until the sixteenth century, gradually came together. These were abstract reasoning, as represented by Euclidean geometry, and experimentation, as represented by the work of the metallurgists who over the generations had improved the methods of winning metals from the ores.[33]

Thus, for Conant:

Science is a dynamic undertaking directed to lowering the degree of empiricism in solving problems; or . . . a process of fabricating a web of interconnected concepts and conceptual schemes arising from experiments and observations and fruitful of further experiments and observations.[34]

The implications here are enormous: historically, many of the great deflection points in human scientific achievement seem to owe their moment to a casually structured interchange between imagination (i.e., manipulation of abstracts), experimentation, and disciplined deduction. In short, the vehicle of science which we seem to recognize in past achievements, far from being abject empiricism, intuitionalism, or unbridled speculation, seems to be the *hypothetico-deductive method*, with empirical validation or invalidation of the conceptual structures being the ultimate arbiter of their scientific acceptability. One must, now, enter an equivocation: the hypotheses which serve as the initial engines in this process need not be rationalistic in origin—indeed, one of the virtues of a concatenative (and empirically disciplined) scientific enterprise is that future hypotheses are able to owe at least some of their substance to the empirically validated hypotheses of the past, and so forth. But it is also of fundamental importance to recognize that the hypothetico-deductive method encourages the postulation, often in the form of broad *heuristics*, of aprioristic masks which can then be used to direct subsequent empirical efforts. Thus, the grand paradigms of the natural and physical sciences (if not those of the social and behavioral sciences) appear to have initially taken the form of heuristics to be employed in a hypothetico-deductive sequence, heuristics which may owe as much of their origin to Neoplatonic imagination or speculation as to the legacy of the experimenters of the past. The demand that constructs eventually be subjected to empirical qualification or testing prevents their becoming transmogrified into the shaggy metaphysics of the scholastics; similarly, the dictate that hypotheses *cum* heuristics need not always be the product of inductive inference prevents science from a priori denying imagination or subjective insight any entrance.

In this sense, we must seriously call into question the somewhat formalistic distinctions made between deductive and inductive inference, between observation and cognitive ordering. No one has seen this more clearly than Errol Harris in his role as philosopher of science:

> Theoretical conceptions permeate the entire process of thinking, and facts always involve interpretation, so that no sharp distinction can be drawn between theory and observation. . . . This independence of observation and theory underlies 'deduction from phenomena,' and gives the lie to any rigid separation of inductive from deductive inference.[35]

This thesis came largely from a study of the *modus operandi* of men like Kepler, Newton, and Harvey. But we can support Harris's contentions with a few remarks about an instance closer to home and our time: the remarkable achievements of Albert Einstein.

Einstein seemed to wear an a priori 'mask' which saw the world as causally consistent and symmetrical, and he once suggested something about 'men dancing to the tune of an invisible piper.'[36] Running deep through his work was this implicit belief in a fundamental order and simplicity in the natural world. Thus, the work of James Clerk Maxwell, which presented as two cases what Einstein chose to view singularly (e.g., the relationship between a conductor and magnet *independent* of the direction of the relationship) led almost directly to the formulation of the special theory of relativity (which was an extension of a kinematic principle suggested by August Foppl: ". . . in the relationship of bodies to one another only relative motion is of importance.") His a priori belief in consistency and symmetry caused him to see a problem where it was not so apparent to scientists not equipped with this particular deductive mask. Thus, he tells us: "The thought that one is dealing here with two fundamentally different cases was, for me, unbearable. The difference was in the choice of reference point." Hence, it was the empirical departure from his valuational position that led Einstein to attack the relativity problem in the first place—and hence to offer resolutions to the paradoxes inherent in classical atomic physics.[37]

But this aprioristic mask, if it provided the leap to his success, also provided him with moments of deep agony and doubt with respect to the problem of quantum physics—a discipline for which he indirectly laid the foundation. Walter Sullivan, noting Einstein's firm belief in 'strict causality,' an indispensable element in his a priori position, suggests the poignancy of the issue of probabilistic phenomena in quoting a passage from a letter written by the physicist in 1924:

The idea that an electron ejected by a light ray can choose of its own free will the moment and direction in which it will fly off, is intolerable to me. If it comes to that, I would rather be a shoemaker or even an employee in a gambling casino than a physicist.[38]

Thus, his denial of the quantum theory was not on the basis of any particular empirical evidence (and, in fact, contrary to much evidence), but because its basic causal contentions were inadmissible or competitive with the *Weltanschauung* he held, specifying strict causality, symmetry, and consistency in all the universe.

The position taken by a physicist not so equipped a priori, and not so many years later, sounds as if it were aimed directly at the almost religious reactionism exhibited in that letter by Einstein. For, in his famous Chicago lectures in 1929, Werner Heisenberg had this to say:

. . . the resolution of the paradoxes of atomic physics can be accomplished only by further renunciation of old and cherished ideas. Most important of these is the idea that natural phenomena obey exact laws—the principle of causality.[39]

Thus, the quantum physics of Heisenberg 'fit' the concatenation of empirical data supporting probabilistic dynamics as a fundamental (and immanent?) causality. And, for physics, deductive inference and rationalistic constructs once more give way to inductively-generated constructs (at least until the next 'crisis,' to borrow a thought from Thomas Kuhn[40]).

Thus, Einstein emerges as a character of incredible complexity in his role as scientist. And it may not be too much to suggest that, as physicists go, Einstein would sit more at the rationalistic end than at the empiricist-positivist (and, of course, the same thing may be said for Crick and Watson as biochemists or molecular biologists, as their discovery of the structure of DNA could hardly have been used as a model of empirical-positivist science[41]). As for Einstein's rationalist bent, we can turn to a contemporary physicist, G. Feinberg:

The notion that philosophy stands behind science in an advisory role is an old one, although it has fallen into disrepute among scientists and philosophers in the twentieth century. One of the roots of the notion is the view that in scientific thought, the mind not only acts to frame concepts relating the raw sense data, but also has a prior role in determining which concepts are considered admissible at all. Einstein was perhaps the most forceful spokesman for this view, for instance in his statement that ". . . the axiomatic basis of theoretical physics cannot be extracted from experience, but must be freely invented."[42]

Feinberg also suggests that there are some dangers associated with this view. Calling on the history of the development of quantum theory, he suggests that: ". . . an effort to impose a priori requirements as to what concepts were admissible would almost certainly have hampered understanding of the phenomena, and the scientists of the time were wise to let the experiments lead the way."[43]

But Feinberg also gives credence to the utility of the rationalistic approach, despite its tendency to develop dogmatic constraints:

There are other times when I believe it is useful to let one's imagination go beyond a consideration of the phenomena being actively studied, and to try to see what sorts of things might in principle exist in the world. This is particularly the case when there is a wealth of experimental material in some area being actively investigated, since that is just the circumstance in which other phenomena may remain undiscovered, or else be ignored. The former seems to have been the case with natural radioactivity, which could have been discovered 50 years earlier, had it been imaginable. The latter happened to the first discovery of parity non-conservation in β-decay, which was reported in 1929, or more than 25 years before it was predicted and rediscovered.[44]

In some ways, then, the history of modern physics is effective evidence for the need to use inductive and deductive modalities in conjunction, and to avoid the excesses that are associated with the exclusive use of one or the other. For, had physics denied Einstein his rationalist contribution, physics would have been denied the theory of relativity; similarly, had Einstein been able to deny the empiricists their license for observation and induction unencumbered by a theoretical constraint as to domain or direction, physics may very well have been denied quantum theory. In reiteration then, going back to Harris's point, the distinction between inductive and deductive modalities can be justified only from a static frame of reference, for in the activist domain of purposive science the distinctions simply do not exist in any meaningful way. And hence we move toward the *empirical* rationale for that ontology we would associate with the general systems theorist.

Particularly, von Bertalanffy's call for the intergrading and integration (or interaction) of *percept* and *concept* gives explicit recognition to the historical evidence for their constancy of association and their complementation in the real business of scientific investigation. In this sense, the general systems theory ontology becomes a distinctly and uniquely triadic one. First, there is the explicit recognition of the role to be played by disciplined deductive inference in the formulation of heuristics which may then be used to guide or direct empirical validation exercises (often in the form of a feedback-loop). There is also the fundamental logic that suggests that simply because a concept developed *ab intra* (with no apparent empirical groundings) there is no guarantee that it will *not* be real—i.e., there is always the possibility that a congery of man's mind might parallel a congery of nature. Hence, the concept has the same a priori right to be considered by science as the percept.

The second leg of the triad is, of course, resting in the domain of empiricals, populated by percepts. Like the concept, the percept has the potential to at least partially determine the substance of hypotheses; also, for essentially simpler phenomena, reduction and successive integration via inductive inference (i.e., synthesis) may exhaust the Reality of the phenomenon at hand, such that deductive postulations become gratuitous. However, for more complex phenomenon, a Baconian approach is likely to bog down, just as science is effectively stopped by the three-body problem. And the percept has still the third role to play: the role as the *necessity* and *sufficiency* for the validation or invalidation of deductively (or for that matter, inductively) generated hypotheses.

As for the third leg of the triad, this assists in the reality convergence of the ontology in essentially two ways: First, there is the necessity to postulate, at least given the current state of the psychological art, that sense data may not be actual representations of empiricals—in short, that some transformation may take place between perception and assimilation or codification. Secondly, there is the problem of leaving a role for subjectivism, revelation, abject intuitionalism, or normative model building. The first leg of the triad, the concept, must be distinguished from these products, largely in the fact that the conceptual structure envisioned there is a product of disciplined deductive inference or axiomatic-metaphysical reasoning—and may often take the form of analogy-building, retrodeduction, or formal system building. However, rather than use the term 'concept' to refer to the outputs of this third component in the triad, we prefer to use the term *hypostatization*. The difference between the first and third legs of the triad may, then, be roughly likened to the difference between mathematics and science fiction, or perhaps more properly the difference between the constructs of rationalists such as Freud, Weber, or Marx and the parabolic, personalized, and poetized constructs of the Eastern mystics (e.g., Sufis) or the European contemplatives such as Law and Frank.

In summary, the general systems theory ontology gives full scope to the range of reality postulations, allowing cognitive-metaphysical constructs, empiricals, and hypostatizations a full reign in the business of gradually creating a posteriori order out of a priori chaos. Overall, then, the general systems theory triadic ontology is primarily unique because it is the *least restrictive and the least dogmatic*. And it must remain so until the questions with which we began our work here have been adequately answered, if ever.

As a closing note, we have tried to justify and expand on the Bertalanffian precept with which we began: the dictate for collapsing the barriers which have so long separated percept and concept. We have also tried to show that the rationalistic, empiricist-positivist and phenomenologistic positions have no logical or empirical (historical) claim to be *the* primary engine of positive, purposive science. Rather, they tend, in practice, to take the form of axiological predicates (or, to avoid euphemisms, *biases*) which exert a troublesome, debilitating but usually transparent or indirect influence on scientific en-

85

terprise. In effect, as ontologies, they give rise to epistemological platforms which are, to employ Eddington's terms: ". . . predispositions inseparable from consciousness" which have an actual if indeterminable effect on scientific analysis.[45] However, they are not inherently transparent or indeterminable. Once we recognize that they exist in principle, we can begin to search for their effects in the empirical domain (as we just did with the brief portrait of Einstein's approach). And while this task is important in the natural and physical sciences, it is absolutely vital within the social and behavioral sciences, where axiological first-premises tend to take on reific implications in that they serve to direct or support policy decisions which, in their turn, tend to introduce alterations into the real world. It is perhaps high time that we, as scientists directly affecting the welfare of society and its constituents, recall this statement by Dewey: ". . . knowledge, mind and meaning are part of the same world they have to do with, and are to be studied in the same empirical spirit that animates the natural sciences."[46]

All we can add to this is the note that, from the general systems theory perspective, these phenomena are highly unlikely to yield their secrets to empiricism unmediated by ordering concepts in the form of broad heuristics. And, by way of a hopefully seminal speculation, we might suggest that *any* Reality that science is able to capture is most likely to be that found *in the nexus of convergence between successively more general inductions and successively more specific deductions*. In summary, it is precisely this possibility which the triadic ontology of general systems theory entails, and precisely this possibility which the diadic ontologies we have discussed obviate by their inherent constraints.

At any rate, we shall be revisiting some of the concepts brought out here in subsequent discussions, for they have a critical impact on some of the key issues remaining to us: (a) Are there real and significant differences between the subjects of the human sciences and those of the natural and physical sciences? (b) Do these differences warrant the methodological and procedural disparities we find? (c) How can we take the ontological and epistemological implications of general systems theory and use them in the establishment of firm operational criteria? It is issues such as these to which we turn our attention in the coming chapter.

3

General Systems Theory as a Counter to Scientism

Introduction

Ultimately, as we tried to convey in the last chapter, epistemological and ontological predicates show up explicitly (if indirectly) in the nature of the models a scientific discipline encourages or builds. We have already suggested something of the character of the models which the general systems theorist admires. Particularly, in the last chapter, we suggested that the models should be Janus-faced, one eye pointing toward the points of empirical validation of the conceptual premises, the other seeking points of articulation with still higher-order constructs or theoretical structures, i.e., successively more generalized models.

Beyond this, however, we have indirectly suggested at least four other characteristics which general systems theory constructs should exhibit:

- The models should have premises which are *fully explicated*, such that we as completely as possible eliminate any tacit predicates (axiological, ontological, or epistemological).
- The models should, to the extent feasible or economical, treat all determinants of a phenomenon as *endogenous*.
- The logical or substantive connections among the determinants (i.e., state-variables) in our models should be both *eidetic* and *nonelliptical*.
- Our models should, to the extent that the properties of the phenomenon at hand permit, be *nomothetic* in character, but the models should always be *apodictical*.

The contention which we shall seek to make clear in this chapter is simply this: failure to adhere to these criteria leads, through one or another mechanism we shall identify, to instances of scientism: (a) Unwarranted reductionism; (b) Assumptivism; (c) Analogic Invention; and (d) Expediency. These instances of scientism, in their turn, lead to the blunted promises and lack or rectitude which we can so often associate with various areas of the human sciences.

1. The Methodological Equivocations

A popular if somewhat self-serving portrait of scientific thinking is that painted for us by Polanyi:

Scientific tradition derives its capacity for self-renewal from its beliefs in the presence of a hidden reality, of which current science is one aspect, while other aspects of it are to be revealed by future discoveries. Any tradition fostering the progress of thought must have this intention: to teach its current ideas as stages leading on to unknown truths which, when discovered, might dissent from the very teachings which engendered them. Such a tradition ensures the independence of its followers by transmitting the conviction that thought has intrinsic powers, to be invoked in men's minds by intimations of hidden truths. It respects the individual for being capable of such response; for being able to see a problem not visible to others, and to explore it on his own responsibility. Such are the metaphysical grounds of intellectual life in a free, dynamic society: the principles which safeguard intellectual life in such a society.[1]

Now, to the extent that the symptoms of scientism are present in the scientific community, this paradigm of Polanyi's becomes significantly more normative than descriptive. The particular symptoms of scientism which will concern us here are those which result from the *ill-considered* application of any one of the following devices:

- *Reductionism:* the assumption that the truth about an entity can be derived from knowledge acquired about its parts.
- *Analogic Inventions:* the imposition of a conceptual or theoretical generalization on an entity under conditions of causal and morphological disparity.
- *Assumptivism:* where variables are treated exogenously rather

than endogenously, when they nevertheless have inherent determinacy on the behavior of the entity at hand.
· *Expediency:* the application of analytical instruments whose resolution power is not sufficient to capture the real properties of the problem at hand.

Clearly, these artifices of science do not qualify as symptoms of scientism except when their potential is abused. Each has a well-defined and potentially positive role to play. Too often, however, they are used to fabricate information about a phenomenon, to illegitimately plug in the gaps between what we think we should know and what is in fact available to us at a given analytical point in time. Thus, in their way, they are all expedients.

The fascination which these artifices hold for the social and behavioral scientist is both real and compelling. We all would be willing to admit that the single-factor analytical methods of classical physics and their exhortation that all real knowledge must proceed from (a posteriori) controlled experiment are worthy ideals, especially when these ideals do not a priori discount the role of theory and conceptualizing in the sense that we explained in the previous chapter. But we would probably all be equally as willing to admit that the phenomena with which the human sciences are forced to deal simply do not admit as fully to controlled experimentation as did the subjects dealt with by classical physics. In fact, the postulated 'Heisenberg effect' takes on critical importance whenever we try to experiment with the volitional subjects of psychology or political science, etc. And, moreover, one must question on the broadest theoretical level the fundamental utility of any experiments which pretend to full control, for there are so few phenomena we might encounter in the human sciences' domain which admit to laboratory closure without, at the same time, losing their ontological significance via unwarranted reductionism or the introduction of artificial constraints.

In short, as Anatol Rapoport has so correctly suggested,

93

the human scientist is in a *bind*. He must admittedly distrust any conceptual systems whose roots lie other than in the empirical domain, validated by controlled experimentation; and, at the same time, he must admit to both logical and practical constraints on his ability to conduct controlled experiments at any meaningful level. Rapoport then suggests how the guises of scientism come to the apparent rescue and, in the process, elucidates their attraction:

There are two ways out of this bind. One is taken by the positivist social scientists who confine their attention to clearly objective, quantifiable, manipulable data, eschewing whatever concepts cannot be related to such data. Thereby they lay themselves open to accusations of trivializing social science. The other way out is to dissociate "understanding" from predictions and control and to declare that the social sciences are based on epistemological principles different from those of the natural sciences. The proponents of the *Verstehen* school of social science stand in danger of succumbing to the sterile methods of the system-building metaphysicians, who peddled their fantasies as revelations of profound truths about the nature of reality and the universe.[2]

This statement is entirely reminiscent of the one we earlier attributed to Robert Merton, where he deplored the partitioning of sociology into two warring (or rather incommunicado) methodological camps: the grand-theory builders and the testers of limited hypotheses. In psychology we note a similar partitioning: between the essentially axiomatic, exegetical system of deductive inferences known as (Freudian) psychoanalysis, and the experimentally-driven, inductive system known as Watsonian (or Skinnerean) behaviorism. Anthropology, economics, and even the management sciences also exhibit such polarization, with the two partitions serving as axiological magnets attracting new entrants to the fields into one or another camp (where the methodological preferences tend to assume importance as membership criteria). Thus, only being a psychologist is not enough: one must also be a behaviorist or a Freudian, etc. As a consequence, methodological preferences, of whatever origin, tend to become superordinate to the emerging realities of the phenomena to be studied, such that schoolism tends to rule the behavioral and social sciences, with ontology being submerged into irrelevance.[3] As a consequence, our disciplines have seldom been able to resist the siren song of scientism in its various guises. For in the last analysis, scientism is simply a situation where methodological factors displace ontological factors in determining the properties of a discipline.

At any rate, we can only be delighted that *both* the deductive and inductive modalities are consistently represented within the human sciences. But despite this, there has not been any convergence of the type which a Janus-faced concept of reality would demand, an attempt at a serious synthesis of the two modalities in confronting the same phenomenon. Particularly, our model-building efforts in the human sciences seem to fall very short of the kind which Kurt Lewin toward the close of his remarkable career suggested would be most profitable. For to Lewin, a substantive discipline represents a "system of concepts" which permits the treatment of both qualitative and quantitative aspects of a phenomenon within a single framework, which entails both causal and descriptive components, which facilitates both definition and measurement tasks and which, ultimately, allows "both generalization to universal laws and concrete treatment of the individual case."[4]

Despite the obvious appeal of such criteria, they represent to some human scientists an impossible ideal because of the "bind" which Rapoport noted. Clearly Lewin leaves us some latitude here, but there is the fundamental impression that there is a premium to be placed on quantification and precision, two attributes which are likely to be achieved only with great difficulty when the subject at hand is a sociobehavioral system.

The human scientist's contention that his subjects cannot

be meaningfully reduced to terms which will permit precise quantitative treatment is more than just a neurotic rationalization. For one thing, many of the determinants of behavior are recognized as being empirically inaccessible in mechanical terms, except by those scientists adhering to a strict empiricist position (e.g., rat psychologists). Ideational, axiological and conceptual factors determine behavior in the most real way, but cannot be precisely measured, isolated, or manipulated within an artificially closed laboratory context. They will, therefore, not be amenable to treatment by the traditional methods of the physical sciences—not without incurring indictments for scientism.[5] However, these arguments can be made more explicit, for many philosophers of the human sciences have sought to make apparent the critical differences between the essentially mechanical ideal-type associated with classical physics and the essentially organic phenomenal ideal-type which seems to best approximate the subjects of the social and behavioral sciences. We summarize these differences in the following table.

The terms employed in the table have definite mathematical and logical implications for the systems scientist, for they reflect structural or dynamic properties which are correlated with certain distinct forms of system behavior.[6] But the essential implication for all of us is that the entity we encounter which inheres the properties of the mechnical ideal-type will be more analytically tractable than one inhering the properties of the organismic ideal-type. The essential reason for this is that essentially mechanical systems meet the empiricist-positivist criteria most fully, for:

· Their properties are most accessible to direct observation.
· Their properties are most amenable to quantitative or precise qualitative measurement.
· The entity itself, and its several subsystems, are most amenable to manipulation within what amounts to a controlled, laboratory environment.[7]

ATTRIBUTES	MECHANICAL IDEAL TYPE	ORGANIC IDEAL TYPE
(1) Interface conditions:	Usually exists within well-defined, tangible boundaries which may be adjusted endogenously for greater or lesser selectivity with respect to entering or exiting forces.	Highly open with respect to environment and exogenous forces. External determinants which affect the system may, therefore, be too far removed (spatially or temporally) to be analytically observable at any point in time.
(2) Structural characteristics:	Generally has its components arrayed in a neat, observable hierarchy such that properties at one level tend to be extrapolations of properties at other levels, with relationships among the various levels being essentially deterministic.	Parts are not arrayed in a neat, stable hierarchy but stochastically, with the structure and direction of interrelationships generally altering almost constantly in response to localized changes of environmental influences, etc.
(3) Dynamic properties:	Parts are usually highly constrained, having only a limited repertoire of responses permitted them; causal trajectories and paths of interaction are generally fixed, controllable and exclusive; driving forces are generally tangible and measurable.	Parts have potential for inaugurating opportunistic or strategic behavior in response to local parameter changes; causal trajectories may be altered locally and interactions may be equifinal; dynamic (driving) forces may be transparent rather than tangible and manipulable.
(4) Normative analytical properties:	Potentially total: —observability —measurability —manipulability —predictability	At least partial: —empirical inaccessibility —immeasurability —imperfect controllability —unpredictability
(5) Amenability to inference and induction:	Given initial state conditions, future states may be induced with a high probability of accuracy.	Future state conditions cannot be successfully inferred from initial state conditions.

Mechanical entities are also more tractable because they can be meaningfully *reduced* to empirically manageable sub-components and then put back together again without incurring a significant error in system synthesis. This is as true of the carburetor in your automobile as it is of the Apollo space vehicle. But this amenability to reductionism and controlled synthesis does not generally hold for the systems dealt with by the ecologist, the social scientist, or behavioral scientist, nor for many of the more complex subjects now being studied by quantum physicists, biochemists or those engineers engaged with adaptive or heuristic systems. In such systems:

- The parts may alter their behavior endogenously, with respect to local factors of which the system as a whole or the other parts may be unaware.
- The behavior of a part may change according to the nature of its interface at any point in time; thus, for example, a person may behave one way in interaction with one individual and alter his behavior when placed in association with another.
- Finally, the range of interfaces to which each component may be subjected (or elect) is extremely wide in social or animate systems; in essentially mechanical systems, interface options are usually invariant and constrained according to initial design criteria.

These take on an added significance when we adopt the perspective that sees applied science as the vehicle by which man designs *systems* to control, modify, or solve problems which are themselves properly defined as systems. For when we are trying to solve an essentially organic problem by designing a system which will prove both effective and efficient, we face frustration at virtually every turn, beginning with the difficulties in developing an adequately exhaustive model of the problem *cum* system:

A. *Properties Inhibiting Problem Definition:*
   1. The *lack of identifiable, definable boundaries* for the problem (system) due to the constant interchange of forces and material with the environment.

2. This means that *some problem determinants will be unidentifiable* at any point in time, being spatially or temporally removed from the immediate problem vicinity.

3. The *dynamic causal sequences* leading to the problem's state(s) may be expected to be *unallegorizable* because of:

   (a) components' capability for *equifinal*[8] *behavior* (getting to a value-point by different paths).

   (b) *reflexivity*, by which every variable in the problem may legitimately be expected to exert determining force on any other (not hierarchial determinacy or otherwise 'ordered' behavior).

   (c) some of the forces determining the problem state will be *transparent* (we won't know they're there), especially those attitudinal and behavioral determinants which can be only reflectively observed (through their impact on observable parts).

4. The *future state(s) of a problem are largely unpredictable* because of structural/dynamic indeterminacies and because their components (human or social entities) have the capability for initiating *strategic behavior* in response to encroachments by the problem-solving system.[9]

B. *Properties Inhibiting the Design of a Congruent (Effective/Efficient Problem-Solving System:*

   1. Because of the inability to accurately determine the future 'states' the problem-system might evidence, there is a probability that any problem-solving strategy we elect to implement will prove dysfunctional. That is, we may have predicated our treatment on an event which will not in fact occur.

   2. Because of the wide behavioral-reactive repertoire available to components of nonmechanical systems, evidenced especially in their facility for strategic behavior, any therapeutic or control instrument we elect to impose may prove dysfunctional because:

      (a) of the lack of empirically validated correlations between instruments and effects within the context of organic entities.

      (b) of the possibility that instruments employed in an organically structured entity (i.e., a reflexive-recursive system context) may interfere with each other or cause a priori unpredictable events which prove inhibitory to positive, purposive control or change.

3. Because of the possibility that the problem-system we are trying to control or modify and the problem-solving system we have designed will be chronically out-of-phase, largely because state-changes may occur more rapidly in the system we are treating than can be adjusted for in the problem solving system we have inaugurated.[10]

In rather more general terms, Sorokin has walked the same ground, suggesting that:

. . . there is the fundamental difference between sociocultural and physiochemical—or even purely biological—phenomena. It consists in a profound difference between the componential structure of sociocultural phenomena on the one hand, and that of physiochemical and purely biological phenomena on the other. Any empirical sociocultural phenomenon consists of three components: (1) immaterial, spaceless and timeless meanings; (2) material (physiochemical and biological) vehicles that "materialize, externalize, or objectify" the meanings; and (3) human agents that bear, use, and operate the meanings with the help of material vehicles.[11]

More simply, it is the concept of 'meaning' which differentiates sociocultural from other types of phenomena; and for Sorokin (apparently), *meaning* takes the form of the kind of multilayered, distributed *volunteerism* (cf. Piaget) which makes sociocultural systems unamenable to natural science treatments. As he puts it:

Mere mechanical borrowing and importation of principles from the natural sciences cannot serve the purpose of the scientific study of sociocultural phenomena; for when literally transposed from one field to the other, they prove to be inadequate. The invariant result is the distortion of these principles as they are given in the natural sciences; the creation of an amateurish and superfluous pseudophysics, pseudomechanics, pseudomathematics, and pseudobiology, running parallel with the real natural sciences; and, finally, a virtual failure to grasp the essence of sociocultural phenomena in their static and dynamic aspects.[12]

And most of us have heard an even stronger statement to this affect by Max Weber himself:

. . . an 'objective' analysis of cultural events, which proceeds according to the thesis that the ideal science is the reduction of empirical reality to 'laws' is meaningless.[13]

But there are others who see such contentions as the out-of-hand condemnation of social science subjects to the realm of rhetoric and ideographic (as opposed to nomothetic) treatment. Among them, and perhaps most kindly, Gunnar Myrdal has asked that we begin to open up, discipline, and codify the treatment of values as proper phenomena (i.e., axiological predicates). But he provides us with an even more penetrating criticism when he asks that we recall that so many social and behavioral scientists may have been far more motivated by some urge to change or manipulate societal systems than to understand them.[14] And when this occurs, science gives way to scientism, for rhetoric displaces analysis and evangelism displaces objectivity.

However, a more pointed and extreme attack on those trying to isolate methodologically-significant differences between mechanical and organic subjects comes, as expected, from the strict positivist. Following Watson's lead, the social scientist George Lundberg exercises the belief that all ontological significance rests in the empirical domain and that all knowledge is gained through the senses. As such, he tolerates no methodological differences between the human and physical sciences, suggesting that:

All phenomena are different in some respects. *All* of them are similar in one highly vital respect, namely, in that they are all known, if at all, through sense experience conceptualized and organized into the patterns determined by the nature of the human organism as conditioned by all its environments. . . . Are the means by which we know societal phenomena fundamentally different than the means by which we know physical phenomena? If they are not, then it is as irrelevant . . . to enumerate "differences" between "physical" and societal phenomena as it would be to claim that the differences between ants, spiders, and grasshoppers preclude a science of biology.[15]

101

As seems so often to be the case, some sort of compromise appears to be most attractive to those sitting outside the recriminatory schools themselves. In the present case, the compromise would be some sort of methodological platform which acknowledges the role of idiographic methods in treating sociocultural phenomena on their axiological dimension, but which also seeks to inject as much of the empirical preference and quantitative discipline of the positivists as is appropriate. The reader will hopefully recall that it was toward just such a union of percept and concept, toward just such an integration of deductive and inductive modalities, that we argued (with von Bertalanffy's assistance) in the previous chapter.

It is not surprising then that a general systems theorist such as von Bertalanffy would realize the desirability for injecting empirical discipline into the human sciences yet, at the same time, reject the mechanistic approach of the behaviorists as scientistic; nor is it particularly suprising that he would applaud Sorokin's emphasis on meaning and symbols as critical properties of human and social systems, yet at the same time condemn the subjectivism or the idiographic-rhetorical excesses of most grand-theory builders. This last point is particularly significant, for von Bertalanffy (along with most other general systems theorists concerned with social and behavioral phenomena) would stress symbolic and axiological factors as *the* major determinants of sociobehavioral phenomena.[16] Indeed, for von Bertalanffy, it is solely the existence of a symbolic capability which lends man and his systems *purposive behavior* and which enables them to elude explanation in strictly mechanical terms. For the purposes sought may often be predicated on personalized, highly variable, and empirically transparent symbolic instances, instances vastly more complex than the biomorphic postulations of Freud or the mechanomorphic factors of stimulus-response theorists. In fact, if man's mind differs from Skinner's rats and pigeons in that the latter rely less on symbolic inferences as behavioral determinants, then any attempt to make human behavior totally intelligible via experiments on animals may be as gratuitous as attempts by pre-Copernicans to calculate the orbit of the sun around the earth.[17]

But our concern with conceptual or symbolic universes for man and his social systems is simply this: they have as much a priori right to be treated as subjects for serious hypothesis-building as do the abstract entities of Freud or the electro-mechanical analogies of Skinner and Watson. The denial of such engines of behavior would be roughly equivalent to the denial of electromagnetic fields in engineering or the denial of gravity in physics.[18]

The skeptic might here suggest that electromagnetic fields and gravity are *more* amenable to empirical capture than the axiological and symbolic determinants of human behavior, that instrumental extension of the senses through voltometers and pressure gauges, etc., have caused these phenomena to ultimately become accessible to the positivist. But the scientist attempting to deal with behavioral phenomena, and the social phenomena they underlie, also has an instrument at his disposal: the fact of intercommunication on the symbolic-axiological plane. In short, we can communicate in a universal if albeit often subjective manner with our subjects, and in the process attempt to employ a type of 'symbolic' experimentation (i.e., the dialectic) to stand beside the physical experimentations of the chemists and physicists. As Freidrich von Weiser once suggested: "We can observe nature from the outside only, but ourselves also from within. And since we can do it, why should we not make use of it?"[19]

With the natural equivocation that one should not out-of-hand generalize from one's own experience, we also can see no reason for not taking advantage of the ability to communicate with subjects and to use elicited responses as tentative heuristics indicating potentially productive lines of inquiry. One must of course, in such communication, be continuously aware of what Merton called 'manifest' and 'latent' distinctions, and the fact that what is made manifest by a subject need not reflect real (i.e., latent) determinacy. Nevertheless, the analyst dealing with individual human beings, or most collectivities which are empirically accessible, does have a definite advantage in the form of intercommunication and interchangeability of concepts and experiences and symbols. In fact, in a remarkable article called "If Matter Could Talk," Fritz Machlup cites nine points of potentially positive association between the human scientist and his subjects. For the human scientist:

- can feel and think like the men whose actions he investigates;
- can talk with other men, learn about their experiences, thoughts, or feelings, and ascertain that these are similar to his own;
- can listen to verbal communications, or read written communications, among persons whose actions he investigates, or among persons of the same type;
- can receive verbal communications, solicited or unsolicited, directly from the persons, or type of persons, whose actions he investigates;
- can make mental constructs and models of human thinking and acting, and can construct theoretical systems involving relationship among ideal-typical actions, counteractions, and interactions;
- can interpret, with the use of his abstract models and theories, particular (concrete) observations of human conduct;
- can interpret, with the use of his abstract models and theories, particular (concrete) data as results of certain types of action;
- cannot build useful constructs and theories in disregard of constructs and theories formed and communicated by men of the type he observes;
- cannot obtain useful data (i.e., the "givens" he is supposed to explain) except through verbal (and often also numerical) reports from men engaged in the activities he investigates.[20]

These, if nothing else, can serve as methodological mitigants of the inherently greater complexity we are inclined (both logically and empirically) to associate with human phenomena.

In short, then, when approaching the subjects of the social and behavioral sciences, the scientist does well to return once again to the concept of the Janus-faced entity, this time on a slightly different plane. For the systems perspective demands that we allow our subjects a potential for operation on two dimensions, neither of which can be a priori denied without severe danger to the utility of our work. On the one dimension, man must be seen as a biological phenomenon, with certain 'mechanical' attributes inherited through his genetic endowments and his proclivity for conditioned (and hence deterministic) behavior within certain ranges. But, on the other dimension, genetic and conditioning factors are *not sufficient* to account for the range of empirical behaviors we can identify (i.e., creativity; deductive inference; the idiot-savant; prodigal geniuses in music, chess, and mathematics; ideationally-driven behaviors). Hence we are led to postulate a higher-order behavioral engine (presumably located in the neocortex) where conceptual operations take place under a system of factors and relationships which is as yet unknown to empirical science. It is thus that Laszlo, from his systems view of the world, asks us to think of man as "a coordinating interface system in the multilevel hierarchy of nature."[21]

Simply then, to emphasize one dimension of man at the complete expense of the other, and hence to emphasize one set of analytical instruments at the complete expense of the other, is to commit an epistemological sin of the highest magnitude, to transmogrify a scientific discipline into a scientistic vehicle. And just how this transmogrification takes place is best explained by an exploration of the symptoms and guises of reductionism within the human sciences.

2. Reductionism and Reductability

There are, basically, two types of reductionism. The first is the deliberate effort to take apart a macrophenomenon—to reduce it to its most fundamental components—with the full intention of eventually trying to put it together again. The second guise of reductionism is both more subtle and more widespread: this is the a priori (i.e., theoretical) exercise of the assumption that the causal-structural determinants of a macrophenomenon are to be found in some lower-order phenomenon.

As to the first type of reductionism, there are basically three ways in which the reintegration, subsequent to the reduction, may be performed: (a) *Successive induction* or concatenation of inferences, given the assumption of causal (if not structural) symmetry among the various levels of the entity at hand; (b) *Additivity*, where both the structural and causal properties of a whole are simple sums (or, in some cases, simple products) of the causal and structural properties of the entity's several components; (c) *Synthesis*, where successively higher-order system components are *integrated* with lower-order system components and where each order or level of the system may entail essentially different causal and/or structural properties (with the system model emerging from lower to higher orders as a synthesis of properties and relationships).

Whether these techniques are scientific or scientistic depends, of course, on the nature of the phenomenon being treated or reduced. We need hardly suggest that the only entity which will be legitimately amenable to the additivity technique will be the rather rare, fully-segmented system which we mentioned in an earlier chapter, the system whose functional-structural parts are all effective replicates and which is effectively acephalic (i.e., nonhierarchical). As for the technique of successive induction, legitimacy will follow only where the entity at hand is one in which causal and/or structural properties are indeed *symmetrically* distributed across the whole. Formal bureaucracies, for example, tend to be structurally (though not functionally) segmented, such that essentially replicative structural subsystems perform different tasks (e.g., one platoon in an army unit might be pulling guard duty in Washington while another platoon is asked to fight in the jungles somewhere in East Asia). This structural segmentation carries with it an implication on the system control dimension: certain level-independent algorithms may be used to control broad behavior for all units (e.g., the elements of the "codes of conduct" apply throughout all levels of the military structure). Thus, when structural segmentation and level-independent control algorithms are present in a system, some essential characteristics of higher-order units may be inferred from properties of the lower-order units. In short, the reductionist-inductivist modality will have some definitely positive if severely limited roles to play in treating systems which are essentially segmented, especially where inferences from one level to the next may be objectified by establishing statistical indices of probable accuracy.

As for the final modality, systems synthesis, we have the most pessimistic case so far as the potential isomorphy of properties among levels is concerned (on either the structural or functional dimensions). Here, new properties are postulated for different system levels with lower-order and higher-order subsystems all being synthesized within some emerging model (emerging, that is, by the gradual and upward extension of the boundaries of integration).

But systems synthesis can be employed for only the most highly localized (or "closed") systems, for if the modern physical sciences cannot rise above the three-body problem in terms of modeling integrative behavior, we can hardly expect to do well in attempting to develop synthesized models built up from the reduction of systems with more than three state-variables (unless complexities are artificially assumed away). Thus, the reductionist-synthesis model-building modality will find only limited applications in the social and behavioral sciences, with the more complex phenomena being approached through the deductive-holistic modality we outlined in an earlier chapter.

Yet, in its allowance of potential structural-causal differences in the different levels of a system, the reductionist-synthesis approach *does* reflect the logical platform of the general systems theorist and the student of general hierarchies (whereas the additive and successive-induction modalities neither look for or are prepared to deal with these differences). And, once again, in treating complex organic systems (whether they are the stochastic phenomena of modern quantum physics or the axiological phenomena of the social psychologist, etc.), these *differences* become the primary working hypothesis which demarcate scientific from scientistic treatments. Speaking as a hierarchical theorist and epistemologist, Mario Bunge translates this concern into an ontological proposition:

01. Reality (the world) *is a level structure such that every existent belongs to at least one level of that structure.* The thesis that all things group themselves naturally into sectors every one of which emerges out of previous sectors, is just a sweeping generalization of the theory of evolution. . . . It also contradicts every shade of monism, for its asserts the originality, hence the irreducibility, of every new level. In particular, 01 goes against physicalism (mechanism, vulgar materialism), for which everything is physical. . . . But the main level hypothesis contradicts also idealism, for which there exists something (life, soul, spirit, or history) which is altogether different from the physical level. . . . indeed, though the world is not homogeneous, it is not divided into separate and isolated realms either. On the other hand 01 is consistent with *integrated pluralism,* an ontology that proclaims both the diversity and unity of the world."[22]

If nothing else then, such considerations as we have posed, along with the ontological postulate of Bunge, should be enough to cast some serious doubt on the utility of treating reductionism as the sine qua non of science, and perhaps some residual doubt on the advisability of allowing empiricism and inductive inference to remain as the *null hypotheses* of modern systems analysis. But our concern here will become even clearer when we discuss reductionism in its assumptive guise.

So, by way of summary, reductionism and reductionist-driven analytical exercises become legitimate under essentially two contextual conditions: (a) Where our analytical ambition is merely to describe an entity, *not* to predict its behavior (for prediction always demands a consideration of interfaces with other entities, which in turn demands the generation and parametrization of exogenous variables—i.e., a holistic perspective); and (b) Where the entity under treatment is fully segmented, such that some form of rude aggregation serves to make the bridge between higher and lower-order aspects of the entity. When these conditions are not met, we incur an *error in synthesis* which will make predictive-prescriptive allegorizations we might build rather more gratuitous than realistic.

3. Assumptivism and Assumability

Assumptive reductionism is not simply an artifice of practical researchers, as was the first instance of reductionism. Rather, here we find the most deep-seated and subtle compulsion to scientism. The clearest example of what we have in mind here must be the rhetorical-evangelical quality which has come to be associated with behaviorism. Amitai Etzioni, a sociologist, has noted and deplored the tremendous pressures on the social sciences to move toward "psychologistic reductionism and micro-analysis," the hallmarks of modern psychology in the stimulus-response tradition.[23] After suggesting some of the sources of this pressure, Etzioni then goes on to offer the case for a macroscopic (i.e., holistic) platform for sociology:

Our macroscopic approach is based on the non-reductionist position that societal units are fruitfully viewed as having emergent properties. This is not to fall into the opposite trap, to imply that societal behavior can be explained fully by emergent macroscopic variables and theories. . . . The non-reductionist position holds only that: (a) an important segment of the variance in societal behavior can be explained only by macro-factors (and *this* is what is irreducible to other variables), and (b) these factors and the relations among them constitute a distinct sub-theory within the theory of action. . . . There are two different kinds of macroscopic emergent properties which, when ignored, make for two kinds of reductionism: The first reductionism does not separate sociology from psychology and from social psychology, whereas the second fails to distinguish macro- from micro-sociology.[24]

Here, then, we have an indication that some variables are proper subjects for reductionism, some not. And, in the strictest sense, we can interpret Etzioni as issuing a warning that the failure to distinguish between sources of behavior which are possibly ideational and those which are chemophysical or organomechanistic, might very well tend to obscure some portion of 'reality.' More specifically, macrofactors are assumed to be definite determinants which are not accessible analytically by the reduction of individuals to deterministic, stimulus-response mechanisms.

The attempt, then, to condemn man irrevocably to the realm of the indeterminate, denying him any predictability via his biological origins, is to be avoided with the same assiduity with which we would avoid the attempt to transmogrify him into the social science's equivalent of the engineer's finite-state machine. Clearly, existentialistic (and some phenomenologistic) approaches are guilty of the first type of assumptivism: the failure to see any points of analogy among social and physical entities, etc., and therefore to lend man and his systems any element of determinacy whatsoever. Paul Tillich has explained this very well:

> The Existential thinker needs special forms of expression, because personal Existence cannot be expressed in terms of objective experience. So Schelling uses the traditional religious symbols, Kierkegaard uses paradox, irony, and the pseudonym, Nietzsche the oracle, Bergson images and fluid concepts, Heidegger a mixture of psychological and ontological terms, Jaspers uses what he calls "ciphers," and the Religious Socialist uses concepts oscillating between immanence and transcendence. They all wrestle with the problem of personal or "non-objective" thinking and its expression—this is the calamity of the Existential thinker.[25]

When we look for assumptivism in the other direction, the failure to admit man any potential for volunteeristic activity, any potential for absolving himself of genetic or environmentalistic determinations, we move to the other *extremum absurdum*. For example, we have Skinner's condemnation of the concept of autonomous man (i.e., the individual who is free to act "in possibly original ways" and who is to be given credit for his successes and blamed for his failures) as prescientific.[26] For Skinner, the 'scientific' view of man is as an entity whose behavior is entirely determined by exogenous factors (or is so determinable), and he suggests that it is in the *nature* of scientific inquiry that the evidence should shift in favor of the nonautonomous alternative.[27]

But it can only be in the *nature* of scientific inquiry that such a shift should take place if, indeed, human behavior *is* deterministic and *is* driven solely by genetic and conditioning factors. Otherwise, such a shift could only be in the interest of those individual scientists whose career has been devoted to evangelizing the nonautonomous theory or to those scientists whose background in mathematics has been so sparse as to deny them the capability for dealing with inherently stochastic phenomena in 'scientific' ways. But, if we interpret trends in modern physical science correctly (and indeed, the trend in the life sciences such as biochemistry and psychophysics), we can see that new analytical skills have permitted (and in some cases have been motivated by) the tendency to notice new elements of complexity and stochasticity in subjects which were once thought to be deterministic. In short, the 'interests of scientific inquiry' are best served by letting the properties of the subjects under treatment determine what analytical instruments and approaches we employ, whereas efforts to define phenomena in terms congruent with the capabilities of existing analytical instruments are clear-cut instances of scientism.

In the kindest light, then, Skinnerean and Watsonian psychology must be treated as a theory fraught with *syntactical* implications and empirical pretentions but, in the last analysis, predicated on tacit assumptions which lend the construct as a whole an effectively metaphysical quality. These are: (a) The assumption that associates scientific analysis with the methodology of classical physics; (b) The assumption that all empiricals can be explained by the laws of classical physics; (c) The assumption that all human behavior can be reduced to a mechanistic interpretation under which the human mind becomes the electrochemical equivalent of the electromechanical computer; (d) The assumption that scientific progress is identical with the generation of successively more generalizing inductive inferences. And our fundamental complaint is not simply that Skinner and his associates operate under these assumptions, but that these assumptions are *tacit*—unexplicated but effective determinants of their practical researches and results.

As for the *explicit* assumptions under which behaviorism operates (i.e., the denial of consciousness; the assignment of all behavior to instances of classical or operant conditioning),

112

with these we have no epistcmological quarrel *because* they are explicit and have an a priori right to exist within the discipline of psychology and to rise or fall on the basis of evidence, not affection. It is the implicit assumptions, however, operating on the epistemological plane, which breed the symptoms of scientism we associate with behaviorism. For these assumptions insulate the fundamental theory from both experiential and theoretical (i.e., logical) attack. They serve to direct the attention of the investigator toward those entities which will serve the cause of the theory and to direct his attention away from explanatory systems which have an equal right to a priori consideration. In this way, assumptive systems tend to become self-fulfilling prophecies by virtue of the procedural and methodological activities they support. Thus, for example, experiments with maze-running, using rats and pigeons, are guaranteed to produce results which support the trial-and-error, operant conditioning hypothesis because: (a) as Russell once pointed out, even Isaac Newton could not have devised a more efficient algorithm for running mazes than trial-and-error, and (b) the physiognomy of rats and pigeons is such that the ratio of their limbic to cortical brain proportions is effectively lower than that found in man, and it is in the limbic system where sense-dependent learning is seated.

Thus assumptivism acts to transform what are really hypothetical predicates of a theory into effective axioms, without the mediating factor of empirical validation. As such, the first-order (or most fundamental) bases of an assumptive theory remain unsusceptible to analysis, and are seldom even considered except by opponents of a paradigm. Instead, we get vast numbers of scientists employing and testing sub-hypotheses (i.e., second- and third-order derivatives) which reflect only the functional aspects of the paradigm, but which inherit (again tacitly) the artificial and hidden assumptions

113

which constrain the theory itself. In such a situation, the phenomena with which a discipline deals might better be considered straw men than empiricals.

For straw men are themselves products of assumptivism, and they become most debilitating and insidious when the assumptions under which they were constructed are most carefully hidden beneath the elaboration of the derivative portions of a theoretical system. And, not too remarkably, straw men are immune to attack from either evidential or theoretical grounds. Thus the concept of nonautonomous man continues to exist despite evidence of 'active' or unconditioned behavior found, for example, by Bruner[28] and despite evidence that "it is a symptom of mental disease that spontaneity is impaired";[29] and nonautonomous man continues to survive as the axis of modern psychology despite the total inability of the stimulus-response schema to accommodate recognized *qualitative* differences between mechanical and conceptual behaviors. In short, classical and operant conditioning can be shown to be determinants of behavior only to the extent that the analogy between the human brain and the electronic computer can be maintained. And because psychology has been so long ruled by the assumptions of behaviorism, our knowledge about mental phenomena fails precisely at the point where the brain-computer analogy fails —there is simply no evidence which suggests how we can *induce* conceptual behavior from the sense-driven conditioned behavior which occupies our laboratory analyses.[30]

On the logical dimension, the tacit predicates of behaviorism tend to look a lot like the Lamarckian assumptions which have failed to prove out in the laboratory. From the now famous Alpbach conversations, we can extract a paragraph by Piaget which logically diminishes the assumption which assigns responsibility for behavior solely to exogenous factors.

Any biological adaption implies two poles. . . . On the one hand, it is an "accommodation," i.e., (by definition) a temporary or lasting modification of the organism's structures under the influence of external factors. But if it were "nothing but" that, the organism would be solely dependent upon the environment, as Lamarck maintained, and one would only encounter products of accommodation ("accommodats") whose direct heredity (heredity of acquired characteristics) would have to be admitted. Any adaption, even momentary, implies a complementary pole which, in very general terms, could be called the "assimilation" pole, and which has the task of integrating external factors into the organism's structures; this necessarily implies a continuity between earlier and late structures. This is why any reaction or response is the expression of its continuous structuralization due to the organism as much as it is due to the pressures from the stimuli, the environment. This polarity of assimilation and accommodation is found at all levels of cognitive as well as of organic development.[31]

In psychology then, we find numerous instances where the mechanistic world-hypotheses which have ruled so long are being replaced by organismic concepts, which are more responsive to emerging realities than the instrumentalism asked for by Bacon and supported by the Industrial Revolution.[32]

In overview then, this seems to be the crucial issue: Can we ascribe a value to reductionism that is more appealing than the a prioristic ontological proposition which suggests that "reality can be reduced to its observable features and that knowledge must limit itself to transcribing these features"? Or, even more pointedly, can we legitimate the case for declining the naturalistic assumption that "the lowest relevant system description is at the biochemical (or even quantum theoretic) level, and that such a description is effectively attainable and effectively determines all higher-level descriptions"?[33] The answer must be an unqualified yes, at least for those systems where *relevant causality* inheres in higher-order systems and is not a simple projection of lower-order algorithms. Just as modern physicists have gone beyond the a priori assumption of strict causality symmetrically distributed (in the transition from classical to quantum theory), so must social scientists abandon the socioeconomic equivalents of

115

strict causality (i.e., Spencerian mechanics; Hegelian dialectics; Keynesian economic theory) and allow a probabilistic component. In so doing, the resolution power of the instruments we use must be increased, for in assuming away strict causality, we rend the neat causal bridges classical physicists and behaviorists would have us build between lower- and higher-order elements of systems. In other words, in the organic/quantum-theoretic realm, each system level is potentially engined by a unique set of algorithms.

But it is not simply in psychology that the concern with assumptively closed system approaches is being evidenced. In sociology as well, the necessity for fully-explicating assumptions so that they may be treated scientifically is also being broadcast. Listen, for example, to Alvin Gouldner:

> . . . if the formal purpose of sociology is to discover the character of the social world, how can it be based upon prior assumptions concerning this character? Doesn't this smuggle the rabbit into the hat and require that the things sociology discovers about the social world be limited by, or depend upon, what it already assumes about it? In some part this must be true; sociology can do no other. Sociology necessarily operates within the limits of its assumptions. But when it is acting self-consciously, it can at least put these assumptions to the test.[34]

The key phrase here is 'self-consciously,' which we may extend to mean that whatever assumptions are used as predicates of analysis are to be stated explicitly (preferably with some indication of how the adoption of alternative predicates might have affected the outcome of the analysis process). When assumptions are left tacit, it is an instance of scientism.

As we noted in reference to psychology, tacit assumptions have a determinacy of their own, which often works in very subtle ways but *always* works to dilute the utility of a construct. In sociology, however, tacit assumptions usually appear to be associated with 'means' oriented studies which point toward some postulatedly desirable 'end' which the means are to achieve. Thus, when we treat the *explicit* assumption of many industrial sociologists—the assumption that there is a positive and direct relationship between industrial amenities and worker productivity—we may note the absence of an experimentally-determined predication for this assumption. But, as George Strauss has pointed out, the explicit assumption (or hypothesis) may simply be predicated on a tacit assumption that derives from the individual preferences of the social scientist himself. In treating the concept of 'power equalization' as a means to industrial peace and productivity, for example, Strauss notes that:

. . . it is a prescription for management behavior, and implicit in it are strong value judgements. With its strong emphasis on individual dignity, creative freedom, and self-development, this hypothesis bears all the earmarks of its academic origin. . . . Professors place high value on autonomy, inner direction, and the quest for maximum self-development. . . . Most professors are strongly convinced of the righteousness˙ of their Protestant ethic of hard work and see little incongruity in feeling that everyone should feel as they do.[35]

At any rate, the vast majority of the scientistic assumptions we encounter in the literature of the social and behavioral sciences rest in the failure of the author to introduce the equivocation that they are actually hypotheses which he simply chooses to accept rather than test. But when the social scientist, in neglect of Popper's caveat about "embracing creeds,"[36] transforms axiological hypotheses into axioms, then the role of the scientist is displaced by that of the rhetorician. And it is he who, after all, is rather more concerned with persuasion than explanation.

But a far more insidious and at the same time more fascinating guise of assumptivism-*cum*-scientism occurs in those contexts where we encounter parochialism—the division of the academic world into partitions which de facto neglect the interdependency of factors from several or many disciplines.

In this instance, we find purportedly predictive and prescriptive models being built which treat variables from other disciplines exogenously or as 'shock variables.' Where extradisciplinary variables are treated as exogenous events, we allow them determinacy but introduce them in the form of parameters whose values have been assigned nonexperimentally. These then serve as constraints or as operator-antecedents, with the intradisciplinary variables then being assigned values which are empirically determined. In the case of the extradisciplinary variable as a 'shock variable,' it is treated as a residual category to explain any variance between predicted and observed 'events.' In other words, it is an error term which is used to transport the burden of non-alignment between allegory and actuality to some other discipline. Thus, to present a somewhat jaded example, when traditional theories of economic development don't prove out in the field (e.g., where injection of capital does not generate an entrepreneurial supply), the reason is often stated to be the inoperationality and imprecision of the models of the sociologists who are responsible for treating such phenomena as entrepreneurial behavior or industrial incentives, etc.[37]

We can suggest that most scientists are more than willing to enter into an interdisciplinary effort when they can be assured that the variables of their own discipline will be given predominance in terms of determinacy. But it is not 'interdisciplinary' when the variables of the other fellow are introduced exogenously, or where we a priori assume that whatever margin of actuality our model eventually falls short of, it is due to the 'shock' value associated with extradisciplinary factors which have never been made explicit. In this context, it is particularly interesting to note the way in which economists use the concept of 'error' or 'shock' as a way of hedging one's allegorical-predictive bets. Here we will listen to Jacob Marschak, an eminent mathematical economist, and may interpret the word 'structure' as milieu:

It follows that if among the policies considered there are some that involve structural changes, then the choice of the policy best calculated to achieve given ends presupposes knowledge of the structure which has prevailed before. . . . In economics, the conditions that constitute a structure are (1) a set of relations describing human behavior and institutions as well as technological laws and involving, in general, nonobservable random disturbances and nonobservable random errors of measurement; (2) the joint probability distribution of these random quantities.[38]

In another place, he somewhat amplifies this:

Too often economic theory is formulated in terms of exact relations (similar to alleged laws of natural science), with the frustrating consequence that it is always contradicted by facts. If the numerous causes that cannot be accounted for separately are appropriately accounted for in their joint effect as random disturbances or as measurement errors, statistical prediction in a well-defined sense becomes possible.[39]

What this amounts to, in essence, is the urging of economists to index all their predictions with some probability of error, and to advance the proposition that they are 'right' if the actual events which occur fall somewhere within the predefined stochastic range.

Such an admonition may seem incredible unless we remember that the original motto of the Econometric Society was: "Science is Measurement." Thus, all those 'causes' which cannot—or perhaps will not—be accounted for separately are to be lumped together into a collective random disturbance which, in turn, provides the basis for a statistical spread to protect the predictor against any of those neglected factors actually having an impact. Hence the concept of the 'shock' variable legitimates parochialism and insularity, and at the same time it provides a defense for the scientist selling his services to policy- or decision-makers. To this extent, the transition from deterministic to stochastic instruments within economics, unlike that associated with the transfer from classical to quantum physcis, is a statistical artifice rather than a response to substantive sophistication.

In summary then, the scientistic symptoms of assumptivism are generally encountered in the following guises:

- The introduction of variables in an exogenous state which are, in fact, endogenous determinants of a phenomenon.
- The proffering of means to achieve ends whose rectitude has been unvalidated (e.g., hypotheses transformed into imperatives).
- The use of 'shock' models which obscure the fact that *actual* determinants have been transmogrified into random disturbances.

Thus assumptivism can either wish away unmanageable forces, or introduce new ones in an effort to take care of the residual variance in a problem. Within the social sciences, the equivalent to *phlogiston* is found in the postulation of such surrogative concepts as 'n-achievement' or the economist's 'equilibrium' factor, the latter borrowed out of hand from the industrial engineers to account for otherwise empirically inexplicable behavior. More insidiously though, assumptivism has its most direct and menacing effect in the phenomenon of *parochialism*—the separation of the social and behavioral sciences into disciplinary cells which act to artificially partition the subjects they treat.

4. Analogic Inventions: A Mixed Blessing

Analogic inventions usually take the form of ideal-type allegories under which various classes of real-world phenomena are logically subsumed. It is these which Stephan Pepper had in mind when he postulated the concept of the 'world hypothesis,' which Alvin Gouldner described as "primitive presuppositions about the world and everything in it."[40] Clearly an analogic model can serve as a mask which, if nothing else, helps isolate those variables (a priori) which are expected to be most productive of information when subjected to empirical observation. But they also have a debilitating effect when they become rhetorical rather than hypothetical platforms. Thus, the a priori 'order' and fundamental logical appeal of the mechanistic analogy, when properly viewed as a world hypothesis or *Weltanshauung*, can not only serve as a guide for empirical efforts but as a perceptual filter which can exclude data and causal inferences which would be incompatible within the context of the mechanical paradigm itself. On a broader front, the concept of analogic inventions enables us to introduce an important if pessimistic appraisal of scientific endeavor, the now famous position of Thomas Kuhn, here paraphrased by Alan Ryan:

> The life which Kuhn outlines for a paradigm is thus that it is the ideology of the scientific community. In its shadow scientists can develop explanations of a more or less satisfactory kind, but it they cannot challenge save at those times when normality has broken down and crisis occurs. At such a point, scientists behave like citizens of a totalitarian state by shifting their allegiance *en masse* to a new paradigm, so long as that promises a new normality.[41]

Kuhn's paradigm about paradigms enables us to introduce the scientist as something other than a super-human, as receptive to emotional-ideational issues as most other men (if better able to recognize them as such). Irrespective of the rectitude of Kuhn's position when stood against the optimistic paradigms of Popper and Polanyi, we have at hand the theoretical mechanism to give some explanation of how analogic models-*cum*-world hypotheses become translated into evangelical platforms (seemingly most especially when they fall away from the hands of the original inventor and become institutionalized as the axis of a 'school'). In such a case, what begins as an innocent ideal-type often winds up as a postulated 'desirable end' to be imposed on society or science itself. Obliquely, this was what seemed to happen to Maslow's

121

hierarchy-of-needs, to Marx's construct of dialectical materialism, to Parson's 'conservative' organismic model of the social process, just as surely as it happened to Cartesianism and Baconianism within the epistemological domain. Indeed, it seems that the inventors of analogic constructs or world hypotheses are seldom as zealous and sweeping in their defense as the inheritors, which probably was the dying thought of Marx when he breathed: "Je ne suis pas Marxiste." Extending this thought somewhat further, if St. Augustine could have foreseen the secular-technocratic excesses sprung from his concept of the 'limited state,' he would probably have uttered a similar last apology.

But the peril in generating and employing analogic inventions is often offset by their scientific value for, as Beveridge has suggested:

An analogy is a resemblance between the relations of things, rather than between the things themselves. When one perceives the relationship between A and B resembles the relationship between X and Y on one point, and one knows that A is related to B in various other ways, this suggests looking for similar relationships between X and Y. Analogy is very valuable in suggesting clues or hypotheses and in helping us comprehend phenomena and occurrences we cannot see. It is continually used in scientific thought and language but it is as well to keep in mind that analogy can often be quite misleading and of course can never prove anything.[42]

Clearly, there is some evidence that analogic processes have proven productive in the generation of theories or explanatory models. For example, Fourier predicated his theory of the conduction of heat on the existing models of fluid-flows; the wave theory of light prepared by Huygens owed its essence to the existing wave theory of sound. Thus, as Berrien suggests, ". . . the perception of analogies among systems has been a powerful means for stimulating a search for additional similarities and the formulation of principles having wide generalization."[43] Indeed, the late Kenneth Berrien himself took the concept of the biological organism and tried to extend its significance into the social domain, as have Katz and Kahn and other behavioral scientists (though in a perhaps less direct way).[44] And perhaps the most widely read organizational theorist *cum* sociologist, Talcott Parsons, also sought to employ concepts taken from the fields of biology and ecology, though his preoccupation with homeostatic or conservative mechanisms has earned him some criticism from those who want to give endogenous disruptive factors more emphasis.[45]

At any rate, the important point to remember in treating the ontological-epistemological significance of analogy-building processes is this: they are always an instance of deductive inference, for a proper analogy always has pretentions to generality but will not be the kind of generalization which is arrived at by successive inductive inference. That is, the analogy is in the first and last instance a conceptual device designed to call attention to isomorphisms, and to engine the attempt to induce causality specific to a phenomenon by imposing a generalized causality. To this extent, analogy-building is an extension of assumptivism (just as assumptivism was an extension of reductionism). But the fundamental difference between assumptivism and analogy-building is that in the latter case we are imposing an entire 'system' of relations rather than single parametric values or sets of predicates (*cum* first premises). Thus, analogy-building involves downward imputation of causal or structural gestalten, where the gestalten usually take the form of generic ideal-types. Thus, a good example of an analogic process is the now popular attempt to impose cybernetic or servocontrol models on systems of many different types, seeking to make specific system behaviors (totally) intelligible in terms of feedback theory.

Thus it is that we find Forrester purporting to treat urban systems as phenomena driven by "first-order, negative feedback loops," where the system is viewed as "goal-seeking" but equipped with only "one important state-variable."[46] Thus, through the mediation of analogic invention, something of an epistemological miracle has been performed: Forrester can treat complex urban systems in exactly the same way that Skinner treats rats and pigeons—as phenomena whose behavior is predicated on operant conditioning which is, itself, simply an exercise in first-order, negative feedback processing. Consider what might happen to projects such as Forrester's "industrial" and "urban" dynamics if the only model he had to revert to for purposive behavior were one which denied conditioned responses their immanency. In such a case, both human beings and the collectivities they form would have to be considered "dynamic, goal-seeking systems that change their environment and are not merely passive victims of it."[47] And if this is the case, then the simple servoanalogy breaks down, as does our ability to collapse all relevant behavioral determinants into a single, fabricative state-variable.

Thus, the temptation of the analogy, especially the analogy based on cybernetics with its current pervasion of the social and behavioral sciences, carries with it a distinct opportunity for scientism (an opportunity, we might add, which seems seldom foregone). This becomes especially clear when we consider the implicit parallels between behaviorism and cybernetics as intepretive-directive schema; and the effort to employ the analogy indiscriminately becomes of special concern when we realize, along with Katz and Kahn, that:

. . . behaviorism, based as it was on Newtonian mechanics, could not deal adequately with problems of organization and structure. The stimulus-response formula assumed too static, too constant and too atomistic a psychological world. The relationships in a field of forces affecting component elements were ignored.[48]

But the history of the social and behavioral sciences' fascination with analogy-building does not begin with the cybernetic-servo models. Spencer must take perhaps the greatest share of the credit for popularizing analogies, for in 1875 he tried to make the biological organism what the servomechanism is today:

That there is a real analogy between an individual organism and a social organism, becomes undeniable when certain necessities determining structure are seen to govern them in common. Mutual dependence of parts is that which initiates and guides organization of every kind.[49]

In 1958, March and Simon would echo this in a somewhat different context, suggesting that: ". . . the specificity of structure and coordination within organizations—as contrasted with the diffuse and variable relations *among* organizations and among unorganized individuals—marks off the individual organization as a sociological unit comparable in significance to the individual organism in biology."[50] The basic improvement over the Spencerian position is the explicit separation of structure and coordination, on the understanding that structurally-identical systems, coordinated or organized differently in terms of the dynamic constituents, should produce different behaviors. Thus we have a sort of organization-theory corollary to the postulation of Piaget which says, in essence, that structural identity and behavior are mediated by personalized "organizations" of experience and perception.

The Spencerian stance depends upon the temporal differentiation of structure and is hence consistent with evolutionary theory, yet the analogies drawn by Spencer are nowhere near as precise and empirically appealing as those developed by later anthropologists, the most significant of whom is probably Leslie White.[51] And the primary debt we owe to Durkheim is the release from temporal bondage and the postulation of a simple and immanent (temporally and spatially) motivator: *the division of labor*. With his emphasis on 'order' in the present as opposed to the time-dependent emphases inherent in Comptianism and Spencerianism, Durkheim laid the foundations for functionalism. In this sense, Alvin Gould-

125

ner suggests that Durkheim "thus began the consolidation of sociology as a social science of the synchronic present. . . ."[52]

Thus we replace the concept of the emergent organism at all levels of the biosphere (reminiscent, vaguely, of the Aristotelian 'form') with the concept of the functional organism, whose structural components at any given point in time are those which are utilitarian in some respect or other (with some opportunity for 'cultural lag' introducing temporal dysfunctionalities). But the platform of functional analysis we associate with the social and behavioral sciences differs from that of its originators, the biologists. As Ernest Nagel tells us:

. . . successful functional analysis in biology is not contingent upon the prior acceptance of any particular theory of organic processes. In particular, it does not rest upon the far-going assumption that the continued existence and operation of every part is indispensable to the organism, or that the actual behavior of each distinguishable component of an organic system is dependent on the character and mode of behavior of every other component.[53]

Recall that this statement must be a direct criticism of Malinowski's concept of functionalism, for he suggested that: "The functional view of culture insists . . . upon the principle that in every type of civilization, every custom, material object, idea and belief fulfils some vital function, has some task to accomplish, represents an indispensable part within a working whole."[54] Thus functionalism in the social sciences injected (a priori) a surrogate of the basic utilitarian world hypothesis, that everything which does exist (or should exist) must serve some ostensible and valuable 'function.'

In overview then, it is not too excessive to suggest that the history of the social and behavioral sciences may be 'reduced' to the history of the phasing in and out of certain analogic engines, all largely deductive in origin and all having at least some pretensions as purveyors of universals. Yet not all employment of analogies has been scientistic, for there are two conditions under which the analogy becomes both legitimate and scientifically productive: First, there is the case where structural isomorphisms among phenomena *are* indeed accompanied by causal isomorphisms; second, under the situation where the analogy is simply used as a working hypothesis to be empirically investigated (or as a tentative heuristic to guide analysis) and *not* as a platform from which specific prescriptions are to be deduced. It is this last point which is particularly important within the social and behavioral sciences, for analogies have tended to be taken up as ready-made conceptual envelopes whose utilization has often been both opportunistic and procrustean. For being conceptual devices of both academic *and* operational significance, the same confusion is bound to attend their employment as attends the use of conceptual principles in general. As the great Dutch sociologist, van Nieuwenhuijze, once put it:

The social sciences nowadays labour under nothing so much as under the concepts of which they are made up. We speak of universals as general operational principles, but it seldom becomes clear whether we mean principles according to which we (subject) operate in society, deal with the facts of social life (object), or principles observable in socio-cultural reality (of which the observing subject is part) conceived as something in operation.[55]

If we add to this the question of universals (analogic engines) becoming mysteriously transmogrified from hypotheses into actionable prescriptions, we confound the picture still further. Clearly, analogies cannot be used to support 'ends' or evangelical positions; they are properly used in the same way as other conceptual (deductive) products: as masks with which an a priori chaotic situation can be at least partially ordered, subsequent to empirical validation of the causality proposed by the 'mask.' Grand operational principles then, in the hands of the serious scientist, simply become systemic, encompas-

127

sing heuristics whose utility lies not in their reification but in their perspective—and here rests the general systems theorists's concern with the analogy-building aspect of his approach.

5. Expediency and Expedition

The last source of scientism, expediency, is usually defended on one or another of the following grounds:

- We need a solution but lack effective instruments to arrive at a solution (a problem which has haunted those who, for example, are trying to develop game-based models to attack sociopsychological behavior).
- We need a solution, but to apply instruments congruent with the real properties (e.g., the true complexity) of the problem would be inefficient or economically infeasible—hence, for example, the tendency of operations research and industrial engineering practitioners to assume linearity in the face of evidently nonlinear phenomena.

So, in the first and probably most common instances of expediency *cum* scientism in the social and behavioral sciences, we employ existing instruments which, by virtue of limits on their ability to handle certain orders of complexity (i.e., to resolve a solution for a given number of state-variables introduced simultaneously), yield solutions which are suboptimal, or, in the worst case, downright erroneous. At any rate, expediency in this guise really amounts to a type of *a posteriori reductionism*.

Yet there is another instance of expediency which must be noted: the tendency to ignore the nomothetic-mathematical potential entirely, and treat *all* aspects of *all* phenomena as if they are entirely unsusceptible to any type of quantification or precise statistical or mathematical treatment. This, essentially, was the complaint which George Lundberg echoed thirty-five years ago;[56] and though his denunciation of any significant grounds for nonmathematical methodology offends many epistemologists, we can see here the evident debilitation associated with efforts which condemn, out of hand, all aspects of societal and psychical phenomena to the realm of rhetoric without regard to their different mathematical properties (or their varying amenability to allegorization in quantitative terms). For example, certain highly institutionalized sociocultural entities are proper subjects for the same type of deterministic treatment as certain types of engineered machines, at least so far as behavioral prediction is concerned.

In treating all aspects of social life in strictly rhetorical terms, ignoring a priori any opportunity for quantification and, hence mathematical and statistical discipline, the social scientist becomes open to the charge of scientism; on the other hand, the use of deterministic natural science instruments to treat obviously stochastic behavioral or social phenomena is also scientistic, as are now viewed the deterministic instruments of classical physics in treating inherently stochastic atomic phenomena.

Let's begin what will be a brief discussion with the obvious entry point: the role of mathematics in the social sciences. First of all mathematics, like chess and music, are products of cognition. The rules of mathematics, its first axioms and its subsequently axiomatic formulations, are a grammar, not any allegorical description of real-world phenomena (except perhaps in the opinion of those mathematicians who, like Descartes, are sure that God was also a mathematician, an applied one).

Within the sciences, mathematics is used primarily for the construction of allegories which are useful in describing, explaining, predicting, or simulating real-world entities. Unless the substantive properties of these entities are known, mathematical models cannot be constructed, any more than we could hope to construct a physical model of something we have never seen. Thus, the quality of a mathematical model-*cum*-allegory depends on two things:

- the amount of information available to the model-builder about the phenomenon at hand.
- the extent to which elements available within the instrumental arsenal of mathematics are sufficient to capture and order that information.

Hence, Newton had to invent calculus to formulate his laws of motion; von Neumann and Morgenstern had to invent game models to treat certain aspects of phenomena inadmissible within existing instrumental frameworks; the operations researchers had to invent linear-programming to handle theretofore unmanageable problems during World War II. In virtually all cases, additions to the instrumental arsenal of mathematics have come from conceptual rather than empirical or inductive modalities; often in response to some problem which only became evident through empirical examination, however.

The mathematics needed for a simple negative-feedback model or a stimulus-response model exist already. But, as we credited Skinner with suggesting, the 'autonomous' man of Maslow or Piaget or von Bertalanffy or Frank George is another matter altogether, and when his behavioral latitude extends further into inherent stochasticity than that probabilistic latitude associated with quantum theoretic constructs, we find we don't have adequate mathematical instruments. Thus, if we wished to model, say, a dramaturgical situation via Goffman, or an opportunistic entity engaged in an interchange with another opportunistic entity, we find that the game theory left us by von Neumann and Morgenstern is not mature enough to be imposed on such a situation.[57] To the extent that this is true, we are left with only the discipline of formal logic to ensure the quality of any verbal paradigms we might construct, as logic in its formal guise is simply an attempt to generalize the fundamental propositions of mathematics and extend its constructive implications into areas where the variables to be manipulated are not susceptible to precise quantification.

We will be delving into the availability of mathematical and statistical instruments with respect to various phenomenological classes in a future section, so here it is enough to suggest that the effective limits on the employment of mathematical allegories is reached when:

- the phenomenon under investigation is ill-structured, substantively or causally, such that a mathematical model would be non-ontic despite its elegance.
- the phenomenon at hand is one which is so inherently complex that the level of sophistication required of the mathematical allegory does not yet exist within the arsenal of available instruments.
- data to support a model (i.e., numericalized events) is inaccessible, such that the allegories we might develop as descriptive vehicles can serve no predictive or control function.

As is the case with analogic or aprioristic constructs in general, even admittedly simplistic mathematical and statistical allegories can be employed in a purely heuristic role. In this case, they are merely employed to discipline certain logically (inductively or deductively) probable associational or causal sequences, not to generate prescriptions or predictions which are to be put into action in the real-world by consumers of the scientific product (the 'applied science' context, per se).

We can also introduce scientism in another expediential guise: the development of approximate solutions via statistical and mathematical instrumentation. When the results of such a process are stated as being products of an expedient, well and good; in such a case, the events predicted by the analysis process are set within some probabilistic range which reflects the analysts' or scientists' expectations about the error associated with the expedient. Thus in numerical methods processes, we ask that the propagation of error be estimated via

131

Taylor Series to account for successive losses of mathematical precision.[58] In operations research we deflate the probable accuracy of results associated with a linear treatment of non-linear entities by the expected deviation of the synthesis of linearity from that which would have ultimated had we used more sophisticated, complex, and costly nonlinear techniques. *It is scientific*, then, to use analytical expedients when the expected cost of using a strictly congruent instrument is deemed to be a marginally unproductive investment. A good example of this is the use in formative stages of research projects of computer-driven 'graphic approximations' to complex functions, where the generation of more precise, numerical solutions would be too expensive to justify (i.e., with each additional state-variable we add in an optimization problem, for example, the computer calculation time will increase almost logarithmically). In the context just described, however, expedience is perceived as identical to marginally justified expedition.

Now we come to some pretty thorny ground: the question of the legitimacy of the application of any quantitative instruments (e.g., the techniques of the natural sciences) within the realm of the social sciences. The only way out of this wilderness is the hard way.

The first step is to consider what might legitimate the use of *any* analytical instrument in any context, and this must be the utility of its output (or product). But we can approach this concept of utility only by knowing the ends to which the analysis process is directed. If the exercise is one devoted to the production of a purportedly predictive allegory of some phenomenon, then the use of quantitative instruments permits *measurably* greater precision than verbal paradigms— because quantitative instruments can work with numerical parameters (e.g., discrete estimates) whereas verbal paradigms are forced to confine themselves to categorical variables which are, by definition, always *interval estimates*. But obviously, the use of quantitative intruments presupposes the accessibility of quantitative data pertinent to the variables of the formulation we are building, for unless one is content with dimensionless quantities (e.g., surrogates or the assignment of numerical values to qualitative categories), the precision of the output depends on the precision of the input.

Now so far as social and human phenomena are concerned, the legitimacy of numericalized data depends on the epistemological perspective adhered to, for these, as we have seen, have a direct effect on the types of phenomena which will be seen as legitimate subjects for study as well as on the methods which will be employed. Thus the empiricist-positivist will demand that all scientific constructs be cast in quantitative terms just as he demands that all hypotheses be phrased in nomothetic terms. Such demands are entirely consistent with his mechanistic *Weltanschauung* which sees all phenomena as essentially deterministic (once we have access to the properties of the entity and sufficient laboratory time to conduct enough manipulations and observations). Hence his subjects must all be susceptible to treatment by the methods of classical analysis (i.e., where we hold factors constant in the laboratory and measure the effect of a single-state variable which we have isolated at a specific empirical trial). From this perspective then, "science is measurement," and measurement implies reduction of all phenomena to quantitative surrogates which then become the basic units of analysis.

The adherent to the rationalistic or Verstehen position, on the other hand, would suggest that the probability of any meaningful sociobehavioral phenomenon being reducible to quantitative terms is extremely low; so low, in fact, that there is very little pressure in some schools (among some disciplines) for any instruction in mathematics or statistics beyond basic linear algebra and the techniques of simple inference estimation. Thus for Weber, sociology was to be an *interpretive* science, which implies that it is also to be an idiographic as opposed to nomothetic exercise. Thus it is no surprise that the constructs he and his followers developed neither depend on mathematics nor lend themselves to empirical validation in which mathematical or quantitative instruments could be employed.

Yet we can only suggest how much poorer the social and behavioral sciences would be without these interpretive-idiographic schema. So the general systems theorist's interest is not in eliminating insight or imagination (or in omitting those subjects from our repertoire of study which do not lend themselves to mathematical allegorization), but in devising ways in which ideographic and nomothetic aspects can be combined into models which draw the best from both worlds, quantifying those properties which are so susceptible, developing 'eidetic' verbal devices to handle those properties which defy mathematical manipulation. Thus, for example, we may applaud the models developed by Herbert Simon for their strong imaginative-interpretive quality, and at the same time admire the fact that they are constructed in such a way as to permit empirical validation using the tools of modern mathematics and statistics.[59] Thus when the model-builder is familiar with the tools of modern science as well as with the nuances and complexities of his subject, there is every opportunity for him to make his construct *apodictical*, even when the properties of the problem at hand do not permit the construction of *nomothetic* models. Scientific analysis is not an all or nothing affair, with our models either being nomothetic and hence apodictical or idiographic and hence non-apodictical. Such a situation, just as does the tendency for our disciplines to split into methodologically polarized camps, reflects a basic misunderstanding of the nature of apodictical and nomothetic properties. The two are neither identical nor perfectly correlated, for it is entirely possible to have a model which is apodictical (i.e., susceptible to validation) without being nomothetic, just as it is entirely possible to have a nomothetic construct which is nonapodictical (and we don't have to go too far back into the history of the 'exact' physical sciences to find examples of elegantly nomothetic models which nevertheless failed the test of relevance). Given a choice then, it is vastly more important that a model be apodictical in nature than nomothetic, but also important that those aspects of a problem which permit nomothetic treatment be exposed to it.

For there is no God-given law that *all* properties of sociobehavioral phenomena must elude the quantitative grammars of mathematics and statistics, just as there is no absolute correlation between nomothetic constructions and ultimate validity. But it appears that the polarization of our disciplines may very well continue to plague us, despite the efforts of the general systems theorist. But at least part of the blame for this can be removed from the shoulders of the social and behavioral scientist and transferred to the mathematician, for he is often a fellow who forgets that his discipline is a tool for others and not simply a religion for an elite few—and thus he, in turn, is often forgotten by the practicing scientist. Scientism via expediency may thus carry two guises of importance for us: First, there is the failure to inject nomothetic treatment to the extent that the properties of the phenomenon at hand permit; second, there is the failure to perceive the fact that mathematics is of the same world as social and behavioral phenomena, and not inherently superior to them. If in the former instance we find the tendency to condemn our subjects to the rather sterile realm of rhetoric and subjectivity, it is in the latter that we find the tendency to believe that mathematicizing something makes it so.

By way of summary, then, the dictates that the constructs of the social and behavioral sciences be cast in eidetic terms, be apodictical in nature, unaffected by tacit assumptions and as completely as possible comprised of 'endogenous' components are not just products of some idle metaphysic. Rather,

135

they are designed as specific counters to the four symptoms of scientism we identified as being points of debilitation for our disciplines. Yet when even the most skilled epistemologist or most well-versed general systems theorist concerns himself with the practical problems of the social and behavioral scientist, he must recognize that the points of transition between scientific and scientistic practices are often obscured, always precariously permeable. So if nothing else, it is clear from the work that we have done here that rectitude in science (like fortune in love, war, and politics), is often predicated on the tiniest turns of judgment, and that scientific integrity is as elusive as the truth itself.

4

Methodological and Instrumental Implications of General Systems Theory

Introduction

In this final chapter of our work, we want to consider some of the operational implications of general systems theory. We shall be interested in working on two levels. First, we want to say something about what a system analysis process looks like. Second, we want to try to introduce one of the most important aspects of the general systems theory approach: the concept of instrumental and analytical *congruence*.

Instrumental congruence, as presented here, will actually be a logical extension of work we did in the last chapter. Particularly, we shall try to give a relatively formal and coherent framework within which questions about the relevance of certain types of instruments available to the scientist may be answered. Hopefully, we shall see that questions as to differences between sociobehavioral and mechanicophysical phenomena (and issues as to the appropriateness of the instruments of the physical sciences) can be approached in a disciplined, objective manner. And, in the process, we shall perhaps also see some rather specific operational procedures which will allow the social and behavioral scientist to build the types of models which have been postulated as 'superior' to those which emerge under alternative methodological platforms.

Analytical congruence works on another dimension: it simply suggests that we can identify certain methodological 'patterns,' each of which is useful at a given point in the systems analysis process, and each of which is most appropriate in working with phenomena inhering certain ideal-type properties we shall set out. Overall, then, congruence demands that we use a strategic investigatory modality and employ instruments which are appropriate to the emerging properties of the phenomenon at hand—and that we avoid the all too prevalent tendency to establish research and instrumental procedures which exist independent (or prior) to the problem itself.

In short, then, we are here again trying to make clear some practical points which promise to improve the quality of the models we build which, in turn, can do nothing else but enhance the relevance of our disciplines.

1. Parameters of the System Analysis Process

Virtually all science progresses through model building, where the models constructed act as effective allegories of the real-world entities under investigation. That is, models are always approximations of the phenomena which we wish to predict, describe or reconstruct. As we have already seen, the 'quality' of models varies greatly, from the idiographic-rhetorical constructs of the rationalists to the artificially constrained but highly eidetic and nomothetic constructs of the strict positivists and empiricists. Between these two extremes rest the Janus-faced, holistic models of the general systems theorist.

The role of systems analysis, per se, is directly related to the model-building process—in fact, it is integral to it. For systems analysis, in the strictest sense, is simply an information-gathering medium, where the value we may impute to information is found in its ability to reduce the error associated with our models. And this error-reduction may take several forms, depending on the ambitions the analyst has for his model:

· If the ambition is to develop a model which will assist in the prediction of the behavior of some phenomenon, then information has utility to the extent that it increases the probability of identity between events predicted by the model and those real-world events which the system or entity realizes. In short, information is useful in this context when it acts to reduce the variance between predicted and actual outcomes.

· If the ambition is simply to capture the structural properties of a phenomenon (i.e., describe its domain dimension), then information is useful to the extent that it increases morphological correlations between the analytical model and the real-world entity.

· If the ambition of the model builder is to causally reconstruct some historical event, then information has utility in so far as it acts to reduce the variance between effects generated by the model and those which actually occurred.

At this point, the general systems theorist may go to work on the methodological-instrumental plane. Specifically, if model-building is the sine qua non of all science, then we have the prime dimension for a taxonomy which should prove extremely useful.

The taxonomy would, of course, attempt to isolate those properties of real-world systems which, in abstraction, would carry *direct* implications so far as the model-building process is concerned—and which would rise beyond the rather parochialized arguments about the differences between natural and human phenomena which we explored in the previous chapter. In this taxonomy, what would be important would be those properties which will serve to either expedite or complexify the model-building process.

Fortunately, the general systems theorist thoroughly familiar with the methods of science and with fundamental epistemology is in a position to develop such a taxonomy. For what we are really concerned about here is the *analytical tractability* of phenomena, where analytical tractability simply suggests something about that entity's amenability to systems analysis techniques. Consider the chart on page 142. These four abstract ideal-types simply represent *intervals* on a continuum of analytical tractability, such that the deterministic entity will be most amenable to accurate and precise model-building, with the indeterminate case representing the system

Deterministic	Where, for any given set of starting-state conditions, there is one and only one event which may be assigned a significant probability of occurence (i.e., as with the finite-state automata).	
Moderately Stochastic	Where, for any given set of starting-state conditions, a limited number of qualitatively similar events must be assigned significant probabilities of occurence (as with the problem of trying to estimate next-period sales levels for a well-precedented product).	
Severely Stochastic	Where, for any given set of starting-state conditions, a large number of qualitatively different events may be assigned significantly high probabilities of occurence (as in the area of conflict behavior or game-based analyses).	
Indeterminate	Where, for any given set of starting-state conditions, there is <u>no</u> event which can be assigned a significant probability of occurence; thus the high probability that some outcome we have not been able to prespecify will occur (as in extremely long-range forecasting problems).	

Figure 1 ($P(e_i)$): — — — (significance level); $E = (e_1, e_2, \ldots\ldots\ldots e_n)$

Figure 2: $E = (e_1, e_2, \ldots\ldots\ldots e_n)$

Figure 3 ($P(R)$): $E = (e_1, e_2, \ldots\ldots e_n) ;\quad R$ (Where $\underline{R} =$ an unspecific residual) type which will be least amenable to accurate description, prediction or causal reconstruction.

If we may make something of a logical leap, analytical tractability then is dependent upon the extent to which a phenomenon under treatment:

- has all its properties accessible to empirical observation.
- has determinants and attributes which are amenable to precise quantification.
- is admissible to experimental manipulation within what amounts to a closed laboratory context.

As Gardner Murphy has pointed out:

Modern science took shape by observing and describing the world outside us. . . . It became "exact science" when it was grasped that things observed can become things measured. With "accesory sense organs," like microscopes and sound-recording equipment for delicate playback, devices for magnifying physical effects and magnifying our sensitivity to them, we can see into the depths of nature and measure more delicately. When Watson, Crick and Wilkinson had shown, as biochemists, that the structure of the DNA molecule had to be a double-helix, the exquisite X rays of Rosa Franklin gave visual confirmation of the structure. With these observational skills go mathematical skills, such as those used by Einstein in the establishment of relativity theory. Today, experimental and theoretical physics, shot through and through with mathematical sophistication, have convinced us that the "real" world is that *physical* world which observation, measurement and calculation have established![1]

In this way, the criteria for analytical tractability look very much like the properties of those phenomena which have yielded their substance to the physical and natural sciences, largely through the media of reductionism and successive induction (except when one recognizes the tacit but critical role played by deduction, as we tried to point out in an earlier chapter).[2]

The essential point, however, is this: as the phenomenon at hand increasingly departs from the criteria of empirical accessibility, measurability and controllability, it will also move from the deterministic category through the stochastic toward the indeterminate, etc. And one more thing must be clear: No particular discipline and no sector of science has a monopoly on any of these ideal-types. In physics, for example, we run the gamut from simple, deterministic entities of 'mechanical' form to the as yet indeterminate 'meson' or 'quark'; in biochemistry, we have the relatively simple systems of several atoms bound according to Pauling's algorithms through the complex but empirically accessible structures of DNA and RNA to the effectively indeterminate mechanisms by which cognitive process take place; similarly, in sociology, we have at the deterministic extreme the highly institutionalized, effectively algorithmic ritual-driven primitive systems and toward the indeterminate end of the continuum incredibly complex, heterogeneous, and ideationally driven political and social macrosystems. In short, not *all* the phenomena dealt with by the natural and physical sciences are deterministic, nor are *all* the phenomena of the social and behavioral sciences invariably indeterminate.

It will help us here to reflect on another aspect of the model-building process. Basically, systems analysis can proceed on one or all of four analytical levels, just as a *complete* model of a phenomenon will contain information about each of the following phenomenal dimensions:

- The *state-variable* level, which finds us trying to exhaust the array of structural properties (i.e., major qualitative or static aspects) of the entity under investigation.
- The *parametric* level, where we try to assign some specific quantitative or categorical value to the state-variables, with respect to a specific point in time and real space.
- The *relational* level, which asks that we establish the nature of the interrelationships among the state-variables and the direction of influence (as in 'static-comparative' analysis in macroeconomics).
- The *coefficient* level, which involves the assignment of a specific

value to the relational variables which expresses the 'magnitude' of the interrelationships among the state-variables at a specific point in time and space.

The implications here might be clearer if we give a brief example. Consider the following equation: $z = ax + b^n/n - 1$. The 'z' factor may be considered the system-state, whereas the 'x,' 'a,' and 'b' factors are state-variables (i.e., major structural determinants of the system-state). The relational aspects of the model are the implicit multiplication operator between 'a' and 'x,' the implicit exponentiation operator for 'b,' and the slash indicating division along with the subtraction operation in the numerator. The coefficient level problem, then, is found in the necessity to lend the variable 'n' a specific numerical value, whereas the parametric problem is solved when we assign specific point-in-time values to the state-variables. This very simple interpretation may, of course, be extended in much the same way to very complex mathematical formulations, to verbal paradigms or models (when they are properly conceived), or even to physical simulation models. The lesson is simply this: A model's accuracy will depend on the accuracy of the analyses conducted at all four levels. More importantly for our immediate purposes, the concept of the four levels of system analysis gives us an opportunity to redefine the four ideal-types we earlier introduced, and to redefine them in a way which lends them more operational import (see p. 146). It is important to understand that these ideal-types represent conditions of inherency; that is, a system which meets the definitional conditions for one or another of these analytical 'states' will be *inherently* deterministic or *inherently* stochastic, etc. Not all real-world phenomena, then, will ultimately become deterministic if we spend enough time analyzing them, or spend enough money on equipment or data bases designed to capture their properties. No one has seriously yet suggested, for example, that local weather phenomena will eventually become deterministic when we have done enough

SYSTEM TYPE	HISTORICAL PERFORMANCE CHARACTERISTICS	SUBSTANTIVE EXAMPLES
(1) Deterministic	No significant relational or structural changes through time—state alterations are negligible.	Finite-state automata; production functions for automated processes; properties of institutionalized (primitive) socioeconomic systems.
(2) Moderately Stochastic	Some significant changes in coefficient and parameter values, but invariant basic structural and relational properties.	Promotional elasticities associated with established markets and products (and demand parameters); market shares within an oligopolistic industry; input-output ratios for bureaucratic organizations demographic factors in given regions.
(3) Severely Stochastic	Significant relational and some structural (i.e., state-variable) changes through time, where relational changes are reasonably well contained and where structural changes are either periodic (replicative) or are drawn from a limited population of state-variables (determinant array).	Stochastic-state machines; distributions of political offices; meteorological phenomena; athletic events; human relations phenomena; military "games"; labor management confrontations; the problem of induced genetic mutations.
(4) Indeterminate	Significant structural changes through time, such that state-variables or major determinants cannot be preassigned except partially and probabilistically; both causal and structural properties either empirically inaccessible or unallegorizable; state-changes independent of prior states.	The fashion industry's market; artistic and creative enterprise; cosmological and teleological phenomena; opportunistic phenomena (e.g, guerrilla units); heuristic machines.

research to understand their predicates. Rather, the assumption seems to be that, irrespective of the amount of analytical resources expended, weather will always entail a stochastic element. Similarly, the work of Piaget suggests that the element of 'volunteerism' or 'activism' associated with the human intelligence will make efforts to reduce human behavior to determinacy somewhat futile, whereas Skinner would suggest that as soon as enough research on behavioral predicates has been done, behavior will become deterministic.[3]

The Skinnerean position, then, is that human behavior is *effectively* stochastic at this point in time, but not *inherently* so. In other words, it is simply our lack of information which causes behavior to appear nondeterministic, and this lack may eventually be eliminated when enough study and experimentation has been accomplished. Piaget, on the other hand, would suggest that even an infinite expenditure of time and analytical resources cannot make an inherently stochastic process an effectively deterministic one. Thus, in this way at least, the polarity of opinion in psychology may be thought to reflect the polarity of opinion in physics, with the determinism of classical physics arrayed against the inherent stochasticity of the quantum theorists.[1]

To set these arguments into a rational framework for analysis, we must add another dimension to our emerging model of the systems analysis process. Particularly, we must say something about the concept of *learning curves*. The best place to begin is to suggest that there are three stocks of information (or sets) associated with the systems analysis process, as follows:

- The a priori is that stock of information about a problem or decision at hand which exists prior to the initiation of a dedicated, formal analysis process. More simply, when the decision-making or problem-solving process is set in a time framework, the a priori state exists at time-$t_\emptyset$.
- As we begin a formal analysis aimed at observing and measuring

147

properties or parameters of interest, we move into a succession of a posteriori states existing variously at times $t_1 \ldots t_n$. Normatively, we expect the stock of information about some problem or entity at hand to increase as the number of empirical observations on parameters of interest increases.

· Both the a priori and a posteriori states are really approximations of some *real* informational state—an abstract ideal in which is resident a fully exhaustive array of information about the problem or entity under treatment.

The analysis process, as a phenomenon, refers to the transition between an a priori state at time t_ϕ and the a posteriori state at time t_1; it also is the vehicle by which we move from the a posteriori state at time t_1 to a more substantial one at time t_2, and so forth. The first two of the three states, then, are approximations of the third—working approximations. And the generation of the second is the 'output' of the analysis process, per se, and the foundations of any applied science allegories or models.

Now, as to the question of the morphologies of event-oriented probability distributions, consider the following diagrams:

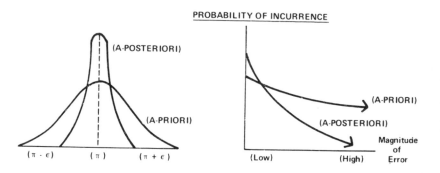

PROBABILITY OF INCURRENCE

The diagram on the left contains two different probability distributions, both built around some variable, π, which may stand for some parameter of interest, perhaps a state-event (e.g., some qualitative outcome) or any of a number of things we may be interested in predicting and/or estimating for purposes of developing some causal allegory.

The A curve represents the a priori probability distribution built around that variable; the B curve the a posteriori distribution. Immediately, we can see that the latter is more favorable than the former, as it involves less variance and, hence, implies a lower probability of incurring an error in estimation or prediction (as the significant probabilities cluster around the true event, π, whereas the distribution of the former entails high probabilities associated with events quite far removed from π).

The diagram on the right reflects the differential probability of predictive and/or descriptive (i.e., causal) error associated with the A and B curves. As the transition from the a priori to the a posteriori proceeds (that is, as the event probability distribution becomes successively more favorable), the probability of error we associate with any allegory we are building decreases. Thus, in an as yet very primitive sense, the function in the diagram to the right, relating the a priori to the a posteriori distributions may be thought of as a *learning curve*, especially when we reassign the dimension for the horizontal axis and introduce the following reconstructions:

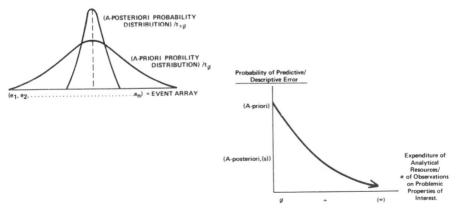

149

The critical proposition indicated in the diagrams is that we normatively expect the probability of error associated with any phenomenon to decrease as the expenditure of analytical resources increases, i.e., we expect the morphological or predictive variance between the real-world entity and the analytical allegory we are building to decrease as the number of observations on properties of interest increases, which subsumes an increase in time and cost of analysis, etc. Thus, as we move from the higher variance a priori probability distribution to the narrower, more favorable a posteriori probability distribution in the diagram on the left, we generate a favorable (declining) learning curve as exhibited in the diagram on the right.

Thus, in surrogate form, we find the imputed value of information in the differences between the decision situation which prevailed at the a priori phase (where no resources and/or time had been spent on formal analysis) and at the a posteriori (where some positive level of expenditure of analytical resources had been reached). Thus, in the most evident sense, the *value of information* is indirectly available to us in terms of the reduction in the feasible solution space associated with some phenomenon under treatment, and the reduction in the feasible solution space is indirectly available to us via the reduction in the expected value of decision error which has occurred in the transformation from the a priori to the a posteriori phase(s). And this reduction in expected value of descriptive and/or predictive error results when we have engined a transformation in event probability distributions such as was indicated in the left-hand diagram, above. There, two things occurred: (1) The extreme departures from π were eliminated (i.e., lent no positive probability of occurrence) and (2) The probability density distribution in the area of π intensified.

Now, when we consider that this 'π' is really an estimator for some 'Π', which represents some true state of the world existing in the inaccessible real-state (e.g., some true parameter; a true state-outcome), and considering the arguments which preceded its introduction, we have the basis for the following *normative* learning curves, 'normative' in that they express roughly the way in which our four phenomenal ideal-types can be expected to behave in the face of formal systems analysis:

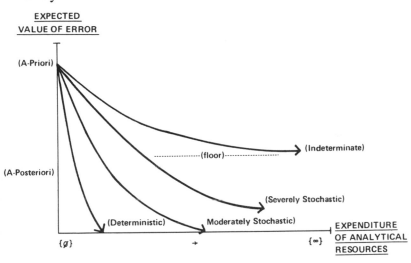

The different learning curves associated with the several ideal-type entities show marked differences in two critical aspects:

(a) The rate of decline in error relative to the base.
(b) The positions of the asymptotes for the learning curves.

The deterministic entity must be taken as the purest embodiment of the properties we earlier associated with the mechanical ideal-type, and normatively we expect its learning curve to drop most rapidly and intercept the zero-error point after the least number of observations or at the lowest expenditure of resources. For the indeterminate system, the surrogate for the organic ideal-type, we notice the least decline per unit of ex-

151

penditure and postulate that there will still remain a significantly positive probability of error associated with our model (i.e., a descriptive and/or predictive allegory) even after a very great number of observations on parameters of interest has been taken (or after a very large expenditure of analytical resources). And, were we speaking about *inherently* stochastic or indeterminate entities rather than *effectively* stochastic and indeterminate ones, we would have to suggest that error variances would exist between our models and the real-world phenomena even after an *infinite* expenditure of analytical resources. In short, there would be no positive probability of reducing these to effectively deterministic entities.[5] At any rate, we can see that we are now in a position to suggest explicitly that the ultimate *quality* of the models we build (where quality indicates the eventual level of structural or dynamic variance between the allegory and the real-world phenomenon) and the *efficiency* of the analysis process (i.e., the rate of reduction in variance between model and real-world phenomenon, relative to expenditures of analytical resources or number of observations) will be determined largely by the inherent properties of the phenomenon at hand.

2. Toward Instrumental Congruence

But it must also be clear that it is not simply the properties of the problem at hand (i.e., the entity under treatment) which determines the quality of the eventual model and the efficiency of the learning or analysis process. There is another critical factor: the suitability of the analytical instruments we elect to employ at the various phases of the systems analysis process. In this regard, we can suggest each of the four analytical ideal-types we have defined will be primarily associated with a particular subset of instruments taken from the general analytical arsenal of the modern sciences. And for those at all familiar with modern quantitative or systems analysis tools and techniques, the (logically) *congruent* associations between phenomenal types and instrument categories will be approximately the following:

INSTRUMENTAL CATEGORY	NATURE OF INFORMATIONAL OUTPUT	MOST CONGRUENT WITH THE FOLLOWING PROBLEMIC TYPE
Optimization	Generates a single solution for any given set of predicates (i.e., linear programming models).	Deterministic
Extrapolative/ Projective	Generates a "range" of possible solutions or a single estimate indexed with a probability of accuracy (i.e., confidence).	Moderately Stochastic
Contingent/ Game-Based	For any given set of informational predicates, will generate an array of cause-effect alternatives.	Severely Stochastic
Metahypothetical/ Heuristic	Generation of heuristics or learning-based paradigms to discipline analysis toward the generation of an adequate set of informational predicates.	Indeterminate

Thus, when faced with an effectively deterministic entity, the instruments with pretentions to optimality are employed (e.g., finite-state system engineering techniques; linear programming; max-min processes). When, however, the entity has not yet permitted the generation of information predicates which will permit absolute accuracy of prediction or description, then we employ the instruments of statistical inference, for example, to generate a *range* of potential parameters, within which we are confident the actual outcome will be contained. In the more serious situation yet, where a large number of different conditions might occur which would, in turn, lead to sets of alternative outcomes, the contingent/game-based models may be brought in to lend some initial discipline to the treatment—especially where our informational predicates at the point in time are so diffuse that significantly different *qualitative* events might emerge (i.e., two individuals in a confrontation may exhibit basically different behavioral modalities, not just differences of intensity of relationship on a single modality). Finally, in response to the worst of all possible worlds, the metahypothetical platform looms most promising; its intention is not to provide any specific directions or solutions, but simply to set the disciplined learning envelope within which a priori indeterminate problems may be initially attacked. In this sense, many of the higher-order, deductively-predicated grand theories of the sciences are metahypotheses whose utility is as heuristics rather than engines for ultimate solutions. In this sense, then, the realization of an effectively optimal learning curve for any given analysis process depends on employing instruments which are congruent (constantly) with the changing analytical properties of the phenomenon at hand. In other words, the central epistemological caveat of Aristotle is given substance and moment, as it was he who warned us in the *Nicomachean Ethics:*

. . . it is the mark of an educated man to look for precision in each class of thing just so far as the nature of the subject admits. It is evidently equally as foolish to accept probable reasoning from a mathematician as to expect from a rhetorician rigorous proofs.

This parsing of the instrumental arsenal of the sciences into instrumental categories which are explicitly correlated with the properties of the four analytical-problemic ideal-types responds to this Aristotelian dictate. This is not immediately apparent, however, until we recognize that the informational output generated by the various instrumental categories differs on a critical dimension: the proportion of the information stock which relies for its substance directly on an empirical data base as opposed to that proportion of the information stock which is nonempirically predicated (or which is deductive or a prioristic in origin).

The mask we must wear to recognize the determinacy of these proportions is as simple as it is critical. First, information, per se, is neither raw data nor unbridled speculation: it is always a product of some portion of a data base pertinent to some phenomenon having been manipulated by some instrument (or by some component of a science's model-base). Think, then, of the information produced during a formal analysis process as a quantum whose origins are jointly rooted in a data base and a model base. Now, from the work we have already done, we know that the reliance on a data base will be greatest for output from instruments associated with the optimization category, least for the instruments of metahypothetical ambition. Thus, if we look at the generation of information in terms of a production function, the implications we want to associate with our four instrumental categories are illustrated below.

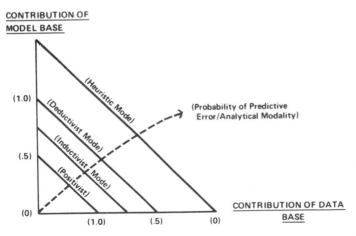

In a sense, the above construct represents a production function for information. It expresses what we already know: that the different instrumental modalities will vary considerably in the composition of the informational quantum they produce; and that the probable accuracy of the allegory we are building in the attempt to predict or describe some phenomenon will decrease as the instrumental base moves from the optimization through the heuristic modality (or, as the proportion of the elements in the allegory generated by heuristic models increases, etc.).

As we move from deterministic toward indeterminate phenomena, the empirically-validated component in the models declines in favor of judgmental or 'deductive' factors, with a corresponding decrease in the morphological correlation between the entity under study and the allegory purporting to describe or predict it. Quite simply, then, we begin to displace fact with opinion and, in the process, incur even greater levels of expected predictive or causal error. Hence, the empirical predication of optimization models is expected to yield (normatively) a lower variance between predicted and actual events than an extrapolative-projective model; these, in turn, are expected to be both more accurate and precise than game-based models. Finally, as we have already explained in earlier chapters, the heuristic or metatheoretical model is expected to serve simply as a tentative device for disciplining subsequent, more detailed learning processes—hence, there is expected to be almost negligible correlation between the models developed under this category and the real-world systems they attempt to treat.

In summary, then, we can suggest that there will be some vector of *congruence* which can be, however grossly, defined between instrumental categories and the various ideal-type systems we established, as below:

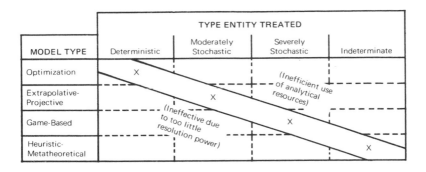

MODEL TYPE	TYPE ENTITY TREATED			
	Deterministic	Moderately Stochastic	Severely Stochastic	Indeterminate
Optimization	X			*(Inefficient use of analytical resources)*
Extrapolative-Projective		X		
Game-Based	*(Ineffective due to too little resolution power)*		X	
Heuristic-Metatheoretical				X

The main diagonal represents this *vector of congruence*, where the given analytical state is associated with that instrumental subset expected to deal most effectively and efficiently with it. We use this term, congruence, to suggest that this instrument class will have the highest 'logical' probability of producing an adequate ultimate error level at the lowest associated expenditure of time and analytical resources. To the right of the main diagonal we strike an expected interface condition called *inefficiency*. The use of this term reflects a probable misallocation of analytical resources here we are using instruments designed to deal with a more complex analytical state than that indicated, and these more powerful instruments will have a higher cost associated with them. Thus, we might possibly employ stochastic tools to solve an essentially deterministic situation, but we will reach any terminal error level at a higher cost than had we used more economical, more expeditious deterministic or optimization instruments.

To the left of the main diagonal falls a more serious situation, here called *ineffectiveness*. The implication is that we are employing analytical instruments which are not powerful enough to resolve the properties of the problem at hand, sug-

157

gesting that whatever eventual error level we do achieve will be unacceptable or considerably higher than necessary. Thus, for example, the economist employing essentially deterministic 'shock models,' in an effort to describe the structural and dynamic properties of a regional economic system, trades off analytical expediency against ultimate rectitude. Any policy decisions made, then, on the basis of that model's output, will inhere a significant probability of being dysfunctional, for they will have incorporated considerable oversimplifications of the real-world problem.

Instrumental congruence, then, is achieved when we employ an instrument which will yield the desired level of information at the lowest associated expenditure of analytical resources—or, more realistically perhaps, will yield a desired quantum of information at a reasonable expenditure level.

There is one more distinction that needs to be made here: between inductively and deductively predicated models. Data, in our framework, is always of empirical origin (i.e., a statistical estimate of some parameter whose actual value rests in the future is not data, per se). But, clearly, the models we may employ are likely to be of two different kinds:

(a) Empirically validated models, which are the product of generalizations whose validity may be determined statistically (as they are usually simple projections developed from empirical data bases, as with the 'laws' of physics or natural growth curves which have been empirically validated for certain populations).

(b) Deductively predicated models, whose origin is nonempirical and whose application is thereby frustrated by the potential of a priori error (as in the introduction of the logically appealing but empirically unvalidated concept of the 'kinked' demand curve into a model attempting to predict the behavior of an oligopolistic enterprise).

Thus, the information entered into an allegory attributable to model bases will vary in its precision and reliability to the extent that the models are predominantly of one or another of

158

these types. Now, recalling again the determinacy of the entity itself as far as the quality of information is concerned, we can suggest that the predicates of the information in an allegory associated with the 'mechanical' and 'organic' ideal-types we defined at the beginning of Chapter 3 (See again page 97) will be significantly different, as the following figures illustrate:

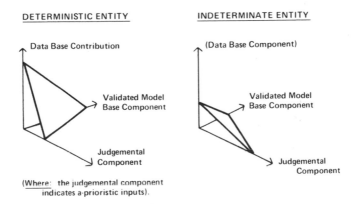

These figures hold constant for the *inherent* mechanism and gestalt (i.e., the inherently deterministic and indeterminate system, respectively). But, where the opportunity exists to gradually transform the effective gestalt into a deterministic entity via the analysis process, we would expect that the morphology of these information maps would alter such that the one on the right would indicate the a priori or initial phases of the analysis, and would ultimately be transformed into the one at the left as the entity moved through the various analytical states. In other words, the model whose predicates are distributed as they are for the mechanistic case, above, will be more reliable and prescriptive than that whose predicates are distributed as in the right-hand figure. And this brings us to another aspect of the general systems theory approach to systems analysis.

3. The Array of Analytical Modalities

The partitioning of the informational substance of an allegory into three predicates or generic antecedents (the empirical data base, the empirically-validated model base and the nonvalidated model base) lends us the perspective we need to generate four basic analytical modalities which have been historically available to the sciences. As the reader might rightly expect, they will differ primarily in their proportional reliance on empiricals as opposed to apriorisms, as follows:

· *Positivistic Modality*: where no logical, mathematical, or statistical model mediates between the data base and the ultimate allegory, such that there is an almost perfect correlation between the components of the original data base and the components of the ultimate allegory. Alternatively, the positivistic modality is indicated by the fact that every element of the allegory is itself deterministic, being assigned no significant probability of departing from the 'assigned value' (e.g., parametric or coefficient).

· *Inductivist Modality*: where the reliance on the original data base is still extremely strong, but where some 'model' has intervened such that the morphological correlation between data base and allegory is dampened—in general, then, inductivist allegories will be extensions or extrapolations of the elements of the data base, such that the parametric or coefficient values assigned the allegory are products of statistical inference, etc.

· *Deductivist Modality*: where the reliance on the empirical data base is rather weak, such that there is little morphological correlation between the components of the ulitmate allegory and the components of the original data base. In other words, under the deductivist modality we introduce significant qualitative changes which mediate between the data base and the allegory.

· *Heuristic Modality*: where there is no data base from which to work which has any relevance for the problem at hand, but only isolated and unintegrated scraps of historical-empirical evidence. Hence, under this modality we deliberately fabricate possible futures and develop initially artificial (i.e., fictional or analytic) frameworks within which a subsequently disciplined learning exercise may take place—where the heuristics employed often owe their origin to idiographic or analogic constructs (i.e., abstract ideal-types).

Now, as we have tried to show in our preliminary arguments, there is no relevance to these modalities that is not explicitly derived from the properties of the problem at hand—the characteristics of the entity we are trying to allegorize. That is, none of the modalities has any inherent ontological superiority from the standpoint of the general systems theorist. For us, the appropriate paradigm for science is that which finds us starting our efforts at understanding a phenomenon with the imposition of metahypotheses to initially discipline the subsequent learning processes. Once the boundaries of the entity have been established and the basic range of properties of interest articulated, we can then turn to the deductivist modality to generate some very broad, unspecific arguments which appear, at that point, to have the highest *logical probability* of being true. These, in turn, give rise to a series of subarguments cast in such a way as to permit empirical treatment (usually within the framework of a Bayesian learning procedure).[6] Their validation or invalidation, in turn, reflects on the validity of the deductively-predicated argument which spawned them, etc. Finally, when the empirical information base has been developed to the extent possible (or economical), the array of optimization instruments may be brought in under the guise of the positivistic modality and as precise and prescriptive as possible a solution to the problem generated.

Thus, *all* the modalities have a positive role to play, and no science may, on the basis of largely *affective* apriorisms, deny them their contribution except so far as they are inappropriate for the phenomenon at its current analytical stage. Thus, the set of major correlations which we want now to begin to defend are the following:

PROBLEMIC TYPE	INSTRUMENT CATEGORY	ANALYTICAL MODALITY
Deterministic	Optimization	Positivistic
Moderately Stochastic	Extrapolative-Projective	Inductivist
Severely Stochastic	Contingent/Game-Based	Deductivist
Indeterminate	Metahypothetical	Heuristic

A. Positivism and the Essential Mechanism

In the extreme case, as we suggested in Chapter 2, phenomenology demands that we generate analytical allegories or phenomenal models which are as fully as possible predicated on a priori unstructured data (and as minimally as possible on any devices called in to either a priori or a posteriori 'order' the data toward any ulteriori ends). Moreover, there is the additional dictate that whatever models we do build restrict their implications as far as possible to a unique, localized phenomenon. In short, we have the hypothesis-free approach to analysis which we earlier associated with phenomenology via Baconism in its extreme form.

This hypothesis-free platform has important if limited applications within the social and behavioral sciences. And we restrict its import to the essentially deterministic (i.e., mechanical) phenomenon for the following reasons:

(a) One of the critical foci of positivism (via phenomenologism) is the adoption of a holistic perspective on entities—the a priori rejection of isomorphisms on the causal (dynamic) dimension of systems obviates attempts at reduction.

(b) However, natural limits on our ability to comprehend wholes in a meaningful way demands that some a priori introduced emphases or 'ordering' precede analysis, except for the simplest, most well-behaved and essentially 'closed' systems.

(c) Thus, the positivist platform, as a potentially productive vehicle for scientific model-building, is restricted largely to those entities which inhere the properties we associated with the essentially deterministic system (i.e., the Type 1 phenomenological ideal-type or the essential mechanism).

This deduced correlation between the positivist approach and the deterministic entity becomes more apparent when we consider just how and where it can be of benefit to the social science community (or scientific enterprise in general).

In terms of the generic analysis process, which finds us successively 'fitting' some analytical allegory to some real-world entity, the hypothesis-free methodology has much to recommend it when the entity can assume only a finite number of 'states,' given any starting-state conditions, such that the ultimately 'true' event can eventually be converged on through permutation-driven *trial-and-error*. Thus, were I trying to reproduce some once-spontaneous chemical reaction (and I knew the constituents and the outcomes but not the 'values' of the constituents, i.e., I knew everything except the parametric values for the state-variables involved in the system), the *most efficient* methodological approach would be to proceed with random alterations in parameters on the assumption that the number of alternatives which must be tried prior to success is finite and that success is an 'identifiable' event *as it is approached*. My a priori hypothesis, in such a case, becomes procedural rather than substantive and the search for a 'solution' becomes identical to the search for a 'cause.' This, then, is the fabled *disciplined* trial and error method of the laboratory scientist trying to reproduce some spontaneously generated event—and this, in the minds of some, is 'science.'

So long as assumptions of determinacy (strict and capturable-reproducible causality) are legitimate, and so long as the number of alternative system configurations which must be tried are adequately finite (or at least presumed so), the

163

need to proceed with experiments predicated on any a priori restrictive hypothesis is obviated. However, these assumptions of finitude and determinacy become successively less valid as the entity, process, or system at hand departs from the criteria of positivism: (a) Observability (b) Measurability (c) Controlled manipulability. Moreover, if there is any probability of the system or entity or process being 'equifinal' (having the capability of arriving at different event-states via different causal trajectories), the hypothesis-free approach becomes fatuous. Thus, while it may be quite successful in the search for, say, an antibiotic to attack a specific pathology in a specific way, or as a method of isolating some effectively-optimal engineering parameter (i.e., interface), it is unlikely to be of much use in attacking a priori ill-structured, organic entities. On the other hand, where the entity at hand is an essentially deterministic mechanical one, the positivist platform is most appropriate. Thus, in its social and behavioral science guise, positivism really becomes disciplined trial-and-error, usually engined by optimization instruments (i.e., finite-state systems analysis algorithms; maximization functions)— and the analyst becomes a transmogrification of Locke's *tabula rasa*, a blank slate to be writ upon solely by empirically-generated sense data (hopefully unadulterated by axiological or exogenous factors). Thus, again, the positivist scheme, with its reliance on nonreductionism and hypothesis-free exercises, must be restricted to those phenomena which are amenable to the following assumptions:

· That the number of trials required to strike the 'optimal' event will be few;
· That the 'optimal' event will be identifiable when approached.

And, in reiteration, this depends on the extent to which the entity under treatment is totally empirically accessible, amenable to precise numericalization or categorization and fully manipulable within an effectively controlled laboratory context. Thus the positivistic modality, in its social science guise, becomes disciplined trial and error, and is appropriate only when we can expect to isolate an optimal event (i.e., eliminate morphological or causal variance between the allegory and the real-world entity) within an extremely limited number of iterations or experimental permutations. Within the confines of our major paradigm, then, the positivist modality makes maximum use of, and demands as a requisite, an extremely accurate, complete, and relevant empirical data base. Thus, for all but the simplest entities, the positivist modality will only become appropriate at the last stages of a systems analysis exercise, when other modalities have effectively removed the entity from the stochastic or indeterminate categories.

This hypothesis-free approach, then, may be thought to sit at the simplest and most 'positive' extreme of the continuum of analytical modalities we will be building—and is congruent with only the most simple kind of analytical problem. The obvious if normative concept we want to gel here then is this: The analytical modality employed in any scientific endeavor will itself increase in procedural and substantive complexity as the problem at hand departs from the properties of finitude (causal) and the properties associated with strict positivism (observability; measurability; manipulability).

B. Inductivism and Moderate Stochasticity

Those adhering to the inductivist modality (which is by far the most prevalent engine of the physical and natural sciences, and increasingly popular among the social sciences) operate under a set of assumptions similar to these:

· There is, first of all, the ontological assumption that 'knowledge' is restricted to the domain of empiricals.
· Second, all empiricals are proper subjects for study.
· This suggests that, for more complex and ramifying phenomena, reductionism is a methodological *sine qua non*.
· However, the concept of strict causality (i.e., Plotinian deter-

165

minism) legitimates reductionism under the assumption that behavior of wholes is inducible from the aggregated behavior of
parts.
· Therefore, the scientific method, per se, should consist of the
empirical observation of entities in a controlled environment (reduced, if necessary, to manageable subphenomena) and the
ordering of these observations into nomothetic propositions in
the form of inductive inferences—where the purpose is identification of generalizations from specifics to the extent that empirical analyses support them.[7]

Thus, the ambitions of the inductivists are both more august
and more portentous than those of the positivists—and their
ultimate target is the instance where a gradually concatenative science will be able to translate successively more chaos
into some sort of deterministic order.

The universalistic impulse of the inductivists makes itself
felt in the logical conclusion of their efforts: the generation of
successively more sweeping generalizations, whose predicates
are founded in objectively (or statistically) determined results
of empirical analyses—not predicated on the artificial axioms
of a deductive structure such as that of mathematics, chess or
music. And the scope of such a scientific edifice, its range of
rationalizing influence, is determined ultimately only by the
limits of the imagination of the hypothesis constructors.

For our immediate purposes, the relationship between the
inductive engine and the matter of competition between data
and model bases is the critical focus. Initially, the resting of
ontological significances in the empirical domain, coupled
with the ambition (indeed, the imperative) to develop generalizations, makes the quality of inductive constructs dependent fundamentally on the rectitude of the empirical data
bases science has assembled. Now, the rectitude of the empirical data base associated with a phenomenon is some joint
product of its relevance, accuracy, and completeness; and
these dimensions of the data base are perhaps best explained

in accordance with properties which tend to make statistical inferences themselves more accurate.[8]

In a broad sense, then, the inductivist is one whose allegories are compounded predominately of either of the following:

- The application of statistical, mathematical or verbal paradigms to a data base, providing that the rectitude of the models employed has been empirically validated.
- The provision of a parameter or coefficient in an allegory where the specific values assigned are products of formal statistical measurement (i.e., regulated by the laws of statistical inference such that the estimates employed may be assigned objectively determined indices of accuracy—levels of confidence).

In this way, the models of the inductivists become allegories whose roots are empirical and whose accuracy is probabilistic —but measurably so.

Thus, inductively predicated allegories express probabilistic engination, such that an allegory may predict (or attempt to reconstruct) a phenomenon's behavior under the assumption that it will behave according to certain empirically-generated generalizations with some significant probability. The data component is very strong, but the actual structure and substance of the allegory owes something to statistical, mathematical, or logical mediation, such that the morphology of the model or allegory need not be isomorphic with respect to the properties of the phenomenon being treated (but it must be a morphological extension or extrapolation). And, to a limited extent, some variables will be exogenous by way of entering certain contextual assumptions, but these too will be calculable, formally-derived extensions of empirically accessed phenomena. Thus, a model imputing a certain behavior to a class of phenomena on the basis of successively more exhaustive analysis of individual members of that phenomenological class is an inductively predicated allegory, and its probable predictive or explanatory reliability is, to a large extent, calculable within statistically generated limits.

It is clear, then, that the validity of any inductively predicated allegory will depend primarily on the quality of the data base from which the inductive inferences were derived (and, where inductively origined models were employed—e.g., the laws of physics—on the extent to which the model-base predicates were empirically validated rather than intuitive in nature). Now, the correlation between the inductive modality and the moderately stochastic phenomenon may be made somewhat explicit when we consider in more detail the determinants of a data base's quality: Accuracy, completeness, and relevance.

As for the aspect of relevance, this simply inolves the consideration of two basic factors:

· The inherent tractability of the entity under treatment.
· The time elapsed between the generation of the data base and its employment as a predicate for inductive inferences to be entered into the model we are constructing.

Hence, the following:

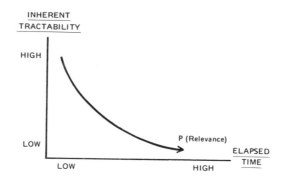

There is a trade-off implicit between the two factors, however, such that if we wished to deal solely with the dynamic dimension of an entity (whose tractability is determined by degree

168

of acceleration and inherent complexity of processual 'states'), we might construct the model as follows:

In the following diagram, data base relevance is assumed equated to probability of error (within the inductive process confines), and the error component is entered along a prime vector as a dependent variable which trisects the model's space from the point of origin outward. Error levels are measured where that vector intersects the planes formed by the three independent variables. Thus, for a given level of acceleration and complexity of change, the probable relevance of the data base increases (error decreases) as the time elapsed between its generation and its employment for decision premises decreases.

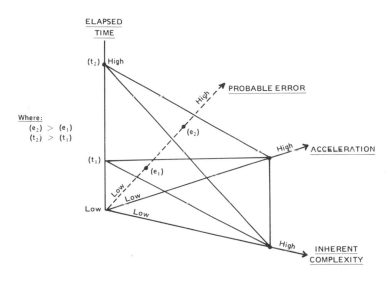

The diagram below indicates the case where we hold the elapsed time variable constant, but vary the other factors from (a_1) to (a_2) and (c_1) to (c_2), respectively. The result is again a case where (e_2) exceeds (e_1) and therefore the probable (or expected) relevance of the data base for Case 1 is greater than that for Case 2:

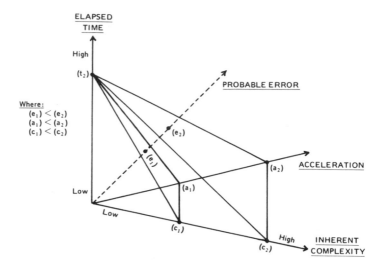

In summary, then, inductively predicated models serve to operate on the relevant components of a data base after the fashion of statistical or mathematical formulae or formalized verbal (e.g., qualitative) paradigms. Thus, to a certain extent, their task is to fill in allegorical gaps (on any level) by extending or extrapolating the substantive components of the data base. Some inductive models may themselves be logical surrogates for empirically observed processes and, as such, are considered empirically validated constructs (e.g., the law of gravity). Others may simply be pseudoaxiomatic, such as the statistical or mathematical instruments which lend an artificial 'order' to previously unordered data.

But this question looms critical and will carry us into the issue of the deductivist modality: At what point should I abandon reliance on the inductivist modality? We can answer this question with respect to suggestions we have made earlier —and repeatedly:

· At the point where the system may be deduced to reside in the severely stochastic or indeterminate portions of the continuum of analytical tractability.

170

- At the point where the historical causal record for the system indicates such severe variance that the probability (statistically assigned) of rectitude or any event extrapolations we make is unsatisfactory or where the replication probability associated with any historically exhibited 'system state' is insignificant.
- At the point where—given any historical performance record associated with previous analytical mode/problem-type correlations—the empirically generated learning curve is found to be unfavorable in terms of the rate at which successive first differences between a posteriori information stocks are declining (such that expected error associated with the model we are building is constant or unresponsive to the aggregate analytical expenditures or concatenative empirical observations).
- At the point where the a priori probability distribution pertaining to statistically or logically probable events is so large as to be unactionable (in terms of the range of events) or where the individual event-probabilities themselves as so diffused as to obviate the predication of action on any one.

The relative appeal of the empirical-inductive-inferential method of analysis then depends on the probable relationship between the observed attributes of the subject system and future or unobserved attributes. Where the probability is low that the causal record of a system will be an accurately guiding precedent for future dynamics (largely dependent upon the variance in the historical record), the inference from past or present to future is unwarranted; where a domain is only partially observable (and that complex, heterogeneous, and clustered), there is little probability that the unobserved will be a comprehensible function of the observed. In short, inference and induction are of declining analytical utility in situations where strong 'organic' properties argue that the future and past, and the seen and unseen, are unlikely to be effectively correlated. Despite the ostensible sophistication of the inferential or inductive techniques which have come riding into organization theory on the back of the computer, there are some systems aspects which are *intractable*, per se. Attempting to coerce inferences or correlations by artificially constraining a subject system, *assuming* away sources of stochasticity, may merely invite the more naive decision-maker to take hold of what looks like a broom but might very well turn out to be an academic mare's tail.

C. Deductivism and Severe Stochasticity

The essential properties of the inductivist modality, in terms of the model-building context were these: (a) The informational content of the algorithm predicated on inductivist techniques will owe the greater proportion of its substance to the data base; (b) The inductivist models will be used, primarily, to effect quantitative extensions or projections of the data base within the context of a single qualitative structure— that is, inductive inference does not introduce properties into the allegory which were effectively unprecedented in the data base itself. But obviously, for entities which approximate the Type 3 entity (the severely stochastic case), the information value to be gained from extending an empirical data base will be scant at best. For such phenomena we will have to make use of the deductive modality, whose models tend to order data via idiographic, empirically unprecedented or *logically probable* (as opposed to statistically or objectively probable) algorithms. Allegorical gaps, then, are filled nonextrapolatively, and, almost always, there is a *qualitative difference* which is introduced in the process. Thus, their utility is in proscribing an order or inference when contextual factors are expected to make extrapolative or pseudoaxiomatic techniques irrelevant, or to operate where allegories or decision premises must be developed in the face of an inadequate historical record (and where time or circumstances do not permit the development of such—as with the nonreproducible 'event'). Thus, deductively-predicated models will be tolerated as the ultimate predicates for an allegory only when inferences are to be made across exceptionally wide tracks of time or space (as with the models of archaeologists or paleontologists) or in the far-future scenario-building activities of long-range forecasters.[9] Or when the entity at hand meets the criteria established for the severely stochastic case.

Particularly, the deductive modality will be characterized by a considerably greater reliance on the model-base than on the data base, such that the allegories developed under its auspices will have an information component which is both less precise and less directive than the data-oriented outputs of allegories engined by inductivist processes. The reason for this is evident: to the extent that the Type 3 system is one whose behavior in the future is not a simple function of its history (or to the extent that the unobserved portion of a severely stochastic stystem's domain is unlikely to be any simple projection of that portion which was or can be accessible to empirical observation), empirical data bases become 'irrelevant' as predicates for predictive or descriptive allegories. And information which cannot be *induced* but which is nevertheless required, must be *deduced*, even if the ultimate rectitude of the allegory is diminished somewhat in the process (as is to be expected when we are dealing with inherently severely stochastic as opposed to inherently moderately stochastic entities).

In short, then, there are three conditions under which the deductivist modality will become most congruent:

· when the phenomenon at hand is *inherently* severely stochastic, such that no amount of historical information collected on observable or observed properties will adequately serve as a basis for structural and/or temporal predictions.
· when the historical record associated with an entity is so incomplete or ill-structured that the future behavior of that entity cannot be *effectively* induced from known or observed properties.
· when there is no host or reference system which can be used to make empirically or inductively predicated inferences about the entity at hand, such that its structural and dynamic 'actualities' are *effectively* unprecedented.

In either of these situations, as has been suggested *ad nauseam*, there is no adequate basis for the positivist or inductivist approaches for purposes of prediction and/or causal explanation. Particularly, an entity falling into these situations must be expected to have the potential to produce different 'states' (e.g., different structural or domain events and different coefficient events) whose predicates are ill-defined or irrelevant. In such a situation, we have no alternative except to try to *deduce*, from whatever sources are available, alternative state or configuration or coefficient events which can be assigned some significant a priori probability of occurrence— where the probabilities are logical or judgmental (i.e., subjective) rather than statistical. Thus we arrive at a distinct and important subcase for the severely stochastic analytical state. Initially, the stochastically 'severe' phenomenon need not always be approached with deductive instruments. Particularly, when the array of alternative events which might occur are replicative or periodic in nature (i.e., precedented), a historically derived frequency distribution might be used to establish the members of the array of alternatives and *induce* their probability of occurrence for the period under prediction (and essentially the same strategy could be applied were our analytical concern cross-sectional or structural rather than behavioral).

More simply, under the normative imputation of severe stochasticity, a system may be expected to exhibit one of several alternative configurations (e.g., engine several different sets of structural properties or several different coefficient-events), but it *is* expected that those which can be predefined will effectively exhaust the possibilities, such that each alternative is assigned some probability of occurrence induced from its frequency or context of occurrence in the past. To the extent, then, that the identification of the several alternative states is predicated on their prior realization according to some evidently operative periodic (e.g., time dependent) or contextual engine, then the prediction exercise (or, for that matter, reconstruction) can be undertaken within the inductive context, possibly assisted by some statistical discipline.

But the equivocation works this way: In cases where there are only a limited number of unprecedented 'state' alternatives which can be assigned any significant logical probability of occurrence on the basis of causal inferences derived from generalistic theoretical or allegorical references, then the severely stochastic phenomenon becomes a subject for deductive inference. In other words, despite the absence of any historically captured periodicity or replicative engination, there exists the possibility to develop an array of a priori 'events' which is expected to exhaust all *logically probable* events. In such a situation, the probabilities of occurrence are subjective or judgmental in origin, yet there is an insignificant a priori probability that some event other than those entered in the deduced array will occur in actuality. It is in this last respect, then, that the deductively-oriented severely stochastic case differs from the indeterminate, where we cannot even deduce an adequately exhaustive event array.

This type of situation occurs most frequently within the confines of the system synthesis problem, where we are trying to ascertain the nature of the interfaces which might emerge between two or more previously unconnected entities. We know enough about them in isolation to suggest some strict range of alternatives which might be 'the' event, and enough in terms of possible interaction characteristics to assign logical probabilities to these various events—even though the problem itself is effectively unprecedented. Indeed, as we shall later try to point out, the option to deduce probable interface events in the face of inadequate or nonexistent historical precedents is a key analytical technique for the social sciences— as it is for systems analysis in general.

In another instance, the deductive option is useful in trying to a priori suggest the range of 'events' which might occur when two previously parochial (e.g., intradisciplinary) models are to be joined—as, for example, in the effort to treat both behavioral and economic variables within the context of an economic development problem. Similarly, a phenomenon for which there are competitive, disparate theoretical or hypothetical stances can also be treated within the deductive confines, such that the competitive theories yield competitive outcomes which are deemed to exhaust the range of 'logically probable' events (which may often be followed by the assignment of some sort of consensus probability index to the alternative theories on the basis of their relative acceptance within the disciplinary ranks, etc. The indeterminate situation, then, differs from the severely stochastic case in that there is no logically probable and exhaustive array of 'state' alternatives which can be developed a priori—hence, as suggested earlier, the overwhelming probability that some totally unexpected and unpredefined event will be the 'real' outcome.

Broadly, then, the deductive modality may be employed to implement any of the following:

· The imposition of a causal order proscribed by a general model on a specific model.
· The imputation of characteristics associated with a general case (e.g., an ideal-type system) to a specific case.
· The imputation of behavioral properties to an entity which are not derived from an empirical examination of the system but drawn, rather, from a theoretical construct of some kind, irrespective of whether or not that construct had inductive-empirical origins.
· The imputation of behaviors which are logically as opposed to statistically probable, such that the origins of the behavioral imputations may be nonempirical *and* nonexperiential.

The deductive potential, then, is in the origination of fabricated, contrived models or hypotheses which tend, more especially, to pinpoint those areas which are not pertinent with respect to any existing generalizations—those areas for which the generalizations are simply not general enough. In this case we fabricate futures in the predictive domain or fabricate causes in the allegorical, with our fabrications generally being composites or modifications of constructs which originated as inductive products. The deductivist then is something of a scavenger, building what often amounts to a work of art out of components which had heretofore never been integrated, leaving some connections unvalidated and often introducing some innovations which even he can hardly rationalize. But in his scavenging and logical machinations lies the future of a truly integrated science and the most sound platform for treating those problems which simply cannot effectively be reduced to isolated, mechanical units of analysis.

To this extent, the deductivist is a fabricator not only of unprecedented futures, drawing on whatever inspiration may exist in terms of unarticulated, abstract correlations and causal sequences, but a fabricator of holistic models from parochial ones—and the efficiency of the applied science front depends, to a great extent, on his success. Thus, while the inductivist may be concerned with developing more and more generalized models all of which, more often than not, are localized because of the reductionist's needs for constraining assumptions, the deductivist is busy trying to identify more and more 'specific' cases in which those generalizations we already have may be applicable—hybridizing where efficiency dictates.

In summary, then, the deductive process is characterized by the generation of predictions and/or causal allegories which are data-independent—derived from some general construct (fabricated or borrowed intact) which purports to treat the class of system to which the entity being treated is ascribed. In a somewhat more specific sense, the deductive analytical mode is used to create specific logical constructs from less specific ones, such that fewer and fewer, more immanent constructs, will be required in aggregate to solve the array of problems facing the as-yet-to-be 'integrated science.' The deductivist synthesizes models of many disciplines by generating interdisciplinary interfaces, under innovative impulse which sends us off not in search of new experiences, but in search of new ways to *create* experience.

Thus, while the basic building blocks of science, barring whatever ethereal influences we might eventually discover, may originate within the confines of the empiricist-positivist camp, the structures they lend themselves to for meaning and purpose are the products of architects whose aim is not so much to codify or articulate experience—or to extend it within given qualitative constraints—but to create new perspectives and ever more penetrating and encompassing logical 'masks' we can wear. And, deliberately or inadvertently, in creating these affective-cognitive vehicles, they may be creating the future itself.

The point to all this is simple: generally, when we encounter an array of categorical (e.g., qualitatively unique, grossly defined) alternative state-events associated with some hypothesis, allegory or theory, etc., we are facing the deductive contribution to the handling of a priori effectively or a posteriori inherently severely stochastic problems. The quest here is for mutually exclusive categories which, when united into an event-array, effectively exhaust the state-events which might occur. Given such an array, we are in a better position to treat the entity at hand, react to it or control it than we would be were such categorizations (e.g., ideal-type constructs) not offered us. In essence, then, the most common task we associate with the deductivist operating within the confines of the human sciences (and the natural as well, to a lesser extent), is not necessarily the generation or fabrication of theories or hypotheses, but the generation of arrays of categorical ideal-types, qualitative event-sets, which are either more precise or more exhaustive than those which preceded them. This is what we will elect to refer to as the defining of qualitative straw men to serve the *stochastic-state process*.

The stochastic-state process is simply the elicitation of such arrays and the assignment of logical probabilities of occurrence to their members as individual alternatives. The purpose of such an exercise is obvious: it serves either of the major aspects of scientific prediction:

- It assists us in *preadapting* to probable alternations in our milieus (which may house systems or entities upon which our own welfare is predicated to some extent).
- It enables us to react with greater precision and expedition to changes which are occuring which, in the absence of the stochastic-state array, might be transparent or so unordered as to inhibit *reaction* (e.g., real-time adaption).

And, just as clearly, the utility of such arrays depends on their exhaustiveness, their precision and—their fewness.

Clearly, the problems of adaption and preadaption represent the highest order concerns of systems—social, economic, political or biological. Somewhat unfairly, it becomes a more critical problem for more complex entities, such that we find ourselves in the interstice between some uncomfortable contentions, as follows:

(a) On the one hand, the most obviously effective means of maintaining ecological congruence or the overall effectiveness of any system operating within a nonplacid ecological context is the prediction of the parametric or state-values which will be assumed by variables or suprasystems affecting the subject system. By the same token, much that we have discussed here suggests the implausibility of such an approach, for the precision and accuracy and probability of occurrence of any event we may specify decreases as we move away from deterministic problem contexts.

(b) In complex contexts, the effectiveness of the prediction-preadaption strategy as a viable analysis-administrative platform can decline, for the only way we can increase the probability of predictive accuracy is to extend the single-event mode to a multiple-event mode, such that *arrays* of events are generated via prediction (i.e., stochastic programming; game theory).

(c) But the more populous the array (the greater the number of events demanding preadaption), the greater the number of preprogrammed contingency responses required. With only limited resources available to any system, the ability to pre-adapt is a priori limited—only a few alternatives may be implemented. Strategically, attempting to prepare to meet too many contingencies lowers the probability of an effective response to the single 'real' event which will occur, just as a general's partitioning of his forces may act to lower the probability that the enemy attack at any one of the predicted locations will be successfully repulsed. Obviously, for any specific situation, preadaption demands a game-based technique which weighs the statistical safety gained from resource partitioning against the dangers inherent in the resultant resource diffusions.

(d) Even in the case of deductively-driven prediction techniques, the most popular of which is scenario-construction, the particular province of men like Daniel Bell and Herman Kahn (scenarists), the probability of the preadaption strategy proving ultimately effective declines as:

- the number of deduced alternative events increases (such that the probability of any one event occurring is a priori deflated).
- the structural or morphological differentiation between the members of the array of alternative events increases (such that preadaption demands significant resource diffusion).

(e) In all then, we find that prediction-preadaption, as an ecological strategy, is most absolutely effective with respect to situations approximating the deterministic or mechanistic ideal-types. For, as soon as we depart from these criteria-sets, we begin to court implausibility, as the solution spaces will have to be increased to the point where the a priori probability distribution of events becomes flat and unactionable (as was the case with the density functions we used to define the indeterminate analytical state). Mathematically, then, when we begin to try to maximize the probability of some true parameter, P, or some true event, E, falling within a predefined range, we expand the feasible solution space to the point where the contingency array becomes unmanageable or impermissible given limited resources—and we deflate the a priori probability of any individual member of the array occurring.

More simply, the greater the array of events we are asked to preprogram responses to, the greater the absolute level of resources required to improve the probability of effectiveness of any partitioned response.

However, within the confines of the severely stochastic analytical domain, the contentions of the scenario-builders and deductivists remain both powerful and increasingly more favorable alternatives to the traditional positivist-inductivist practices. Particularly, so far as the functionality of the stochastic-state approach is concerned, we can cite a principle underlying the brilliant work of Fred Emery looking at "The Next Thirty Years":

In trying to characterize large complex social systems, we are reminded that some behaviors of both organisms and organizations are functions of gross overall characteristics of the system of which they are parts and which constitutes their environment. We can advance our knowledge of these behaviors if we can identify some of the ideal types that characterize the overall environment, as seen from the viewpoint of the generalized part-system relation.[10]

In a somewhat more erudite way than we have done, Emery defends the concept of the generation of those deductive 'masks' associated with the departure from inductive-empirical bases. And, so long as our effort extends to the development of such masks associated with inherently stochastic entities (as opposed to inherently indeterminate entities), the stochastic-state process, with its concentration on predefinition of gross event-alternatives and the provision of contingencies predicated on the occurrence of *all* members of the set, will become increasingly an actionable and beneficial ambition for complex analysis processes undertaken by social systems theorists.

However, to the extent that the entity at hand is not obedient or well-behaved with respect to the severely stochastic criteria—to the extent that a manageable array of alternatives cannot be deduced or induced as an exhaustion of logical or statistically probable events—then we have little recourse except to turn to the heuristic alternative.

It is the employment of the heuristic analytical modality in the face of effective (a priori) or inherent indeterminacy to which we now turn our attention.

D. *The Heuristic Contribution*

As was suggested earlier, the heuristic modality is the one which relies least on the contribution of any data base component—in fact, in their origins, heuristics are largely intuitive and therefore always tentative.

The concept of a 'heuristic,' and various explanations of the utility of heuristic paradigms, have been given by many authors: Newell, Shaw, and Simon; Burstall; George; Samuel Tonge.[11] As a rule, however, the heuristic itself is thought of as an algorithm or paradigm of some kind (George calls them "rules-of-thumb") which attempts to make the solving of a priori ill-structured problems more efficient than would be the case were temporally-sequentially disciplined inductive or deductive processes employed, and more disciplined than were abject intuition or unconstrained trial-and-error opportunism the investigatory predicates. In virtually all cases, however, a heuristic presupposes a gestaltlike, initially holistic view of some phenomenon to be treated, with the often explicit condition that that phenomenon be *a priori indeterminate*, in terms of the definition foundation for the Type 4 entity. In this sense, then, the heuristic becomes a 'learning' instrument which attempts to discipline the process of nonparadigmatic inquiry, that which is not a priori responsive to any of the other modalities we have outlined in this section.

In this sense, many of the higher-order 'paradigms' of the social and behavioral sciences serve as heuristics, a priori 'masks' we can wear to lend some tentative order to otherwise ill-structured (effectively indeterminate) phenomena. As opposed to a theory, per se, which depends for its utility on its apodictical quality (i.e., its ability to be empirically validated), the heuristic makes no pretense to rectitude or allegorical alignment with some specific real-world entity. Rather, it serves merely as a synthetic construct which is usually a combination of two distinct components:

- A *metatheoretical* component which serves to provide an initial and quite broad boundary for analysis. This may, for example, simply involve the loose articulation of a set of variables which we logically feel might act as determinants of the entity under investigation, without any attempt to specify the nature of relationships or restrict subsequent analysis to these variables. On the other hand, this component might involve the postulation of some analogic engine as a first rough approximation to reality or simply involve the imposition of entirely abstract ideal-type implications as a 'template' for the initially a prioristic attack on some problem (e.g., the four phenomenal ideal-types with which we began this chapter serve such a mission). In short, then, for very complex problems or phenomena, the heuristic serves as an analytical structure within which our a priori information may be housed.
- The second component of a proper heuristic is *algorithmic* in nature: it really involves the specification of the 'rules' we will use in disciplining subsequent analysis, setting in the broadest way the criteria of investigation and truth. Thus, for example, a heuristic must specify whether the components of the ultimate allegory can be built up using Bayesian statistics (i.e., conditional probabilities with the weights distributed independent of the order of observations) or according to some sort of weighted scheme (i.e., a Markov process, where latest or most current observations are given relatively more weight than previous observations). To a certain extent, then, this aspect of the heuristic takes on epistemological overtones whereas the first component tends to be substantive in nature.

Thus, a heuristic serves to facilitate and guide research, but is much less directive and specific than would be a deductivist construct which might serve as a predicate in some hypothetico-deductive process. Quite simply, a heuristic would be in existence *prior* to the development of the deductively-generated hypothesis, and would in fact be a prerequisite to its development were the phenomenon at hand indeterminate in its immediate properties.

This brings us to a critical point. Initially, the value of any hypothesis in the positivist, inductivist, and (to a lesser extent) in the deductivist modalities may legitimately be thought to be related to the 'narrowness' of the search and solution space it sets out. Just the opposite may be true of the heuristic: we want it to allow the broadest possible initial range of inquiry, imposing successively more narrow search and solution constraints only as we gradually converge on a solution through one or another of the other modalities. Simply, it is the function of the heuristic to inject manageability into analyses, not constraints. For manageability in the systems analysis process refers to procedural aspects, whereas a priori constraints always imply an origin in the substantive properties of the phenomenon itself. Thus, whenever a construct imposes constraints which cannot be justified by a real or assumptive knowledge of the properties of the phenomenon under treatment, it ceases to be a heuristic and becomes, instead, a theory. It would hence be subject to all the criteria we have laid down for proper theory, criteria which are simply impossible to achieve at the start of an investigation into an unprecedented problem or indeterminate phenomenon. The failure to distinguish between the properties appropriate to a heuristic and those appropriate to theories has led us in the human sciences to the situation where we have very few 'good' instances of either, and many constructs which are neither one or the other and which serve both functions equally poorly.

Thus, the heuristic, when properly employed, enables us to exercise our imagination and speculative abilities without being indicted for scientism. It is thus a vehicle by which we can cover truly vast reaches of academic territory with great efficiency, equipped perhaps with the interest in analogy-building which will allow us to scan for broad patterns of *association* and perhaps generate (or fall into) those instances of *bisociation* which appear to be so necessary for the progress of the sciences.[12] In a more restricted sense, however, it is a device by which we can bring all our past experience to bear on a new problem without causing past experience to be transmogrified into assumptions, per se. In short, the heuristic serves an ordering function, but not a prescriptive one except in the procedural sense.

In summary, then, the heuristic modality simply acts as an entirely *opportunistic* but formally disciplined attempt to impose some manageability on a priori chaotic situations. In the metatheoretical aspect, it will try to set the initial range of inquiry in a way that we can be absolutely certain that the *true* properties of the phenomenon must be contained within this range. Thus, for example, we might approach a highly complex problem about which little is known by first establishing two fabricative (i.e., analytic) hypotheses, the first being the antonym of the other. If we construct them correctly, and if our imagination does not exceed our logic, then these two polar hypotheses should represent the extremes of possibilities, such that the real-world event must fall somewhere within the interval they define (along a dimensional continuum). In its second aspect, the algorithmic, the heuristic simply serves as a residence for our epistemological and methodological skills and sensitivities, where these too take the form of initial working-hypotheses rather than instances of rigid dogma.

4. Some Final Notes

In what we might consider to be the normal course of events, a phenomenon will be expected to respond to our analytical efforts and expenditures by becoming successively more deterministic. Thus, adding a dynamic element to our discussions of the several modalities contained within the general systems theory repertoire, we emerge with the facing illustration where one modality (and, as a consequence, one set of analytical instruments) gradually gives way to another as the analysis proceeds:

Within the framework of this paradigm, assuming that the phenomenon at hand comes to our attention initially as an effectively indeterminate one, the heuristic modality acts to provide us with an envelope of 'pointers' which, in turn, lead to certain broad deductive hypotheses which are self-contained (e.g., not dependent on the substance of the heuristic) and which usually try to structure the several unique system-states which might occur within the severely stochastic context. At this point then, we have at least outlined some possible alternatives. Then, these state issues are resolved by empirical validation of the alternatives until a convergency on one is found. But this presumes that the entity at hand is not *inherently* severely stochastic, for if it were there would be no convergence on a single state, but only the gradual refinement of the probabilities associated with the *several* significant alternatives.

When a single 'state' has been isolated, inductive models may be brought in from the pertinent disciplinary model bases which express the probabilities associated with the structural variables comprising that state assuming some range of parameter values. Eventually, assuming that the entity at hand is inherently deterministic (though it was a priori effectively indeterminate), these probability distributions can be displaced by single parametric estimates which have a significant probability of being the 'only' parametric event to occur with respect to some state-variable. When this occurs, then, deterministic models may be employed which express the conditions for optimization (e.g., predictively), or precisely reconstruct the causal history of the entity.

ANALYTICAL STATE	INDETERMINACY	SEVERE STOCHASTICITY		MODERATE STOCHASTICITY	DETERMINACY
ANALYTICAL MODE	Development of Heuristics	Development of Deductive Hypotheses	Validation of Hypotheses and Refinement	Employment of Inductive Models	Development of Optimization Algorithm

Hence, all four of the modalities outlined have a definite role to play within the general systems theory process, but none of the four modalities can, alone, act to advantage except within certain rather well defined intervals of the problem-solving, systems analysis process.*

And in the framework of the analysis we conducted in this chapter, it does not take too much imagination to see that if one of the modalities is used to the exclusion of the others (as each of the traditional epistemological positions we earlier described would tend to dictate), we can emerge with strikingly different appreciations of any given phenomena, where none of the appreciations can provide anything except preselected, artificially manipulated perceptions. As an alternative to these platforms, however, general systems theory draws on the positive contributions which can be made by each (and indeed which must be allowed to each within the respective spheres of authority which we have defined) and, as a result, allows the social and behavioral scientist to treat the full range of phenomena he must be prepared to deal with. For, if any of the modalities were eliminated, the scientist will lose the instrumental and analytical congruents associated with those entities which fall into the ideal-type category that modality uniquely treats. This might perhaps go a long way toward explaining why *all* sociobehavioral phenomena tend to appear as deterministic to empiricists such as Skinner or Watson, and why *all* sociobehavioral phenomena tend to appear *inherently* indeterminate to idiographers such as Weber or Sorokin.

Thus, if one were to search for a lesson from this survey of general systems theory philosophy it would be this: the ultimate ontological significance of the social and behavioral sciences will be inhibited by adherence to any epistemology or

---

* For a more mathematical treatment of general systems theory as a discerning rather than eclectic approach, see my article: "Beyond Systems Engineering," published in the 1973 edition of the *General Systems Yearbook*.

methodological platform which either dictates or uncon-
sciously accepts a restricted interpretation of reality, and
which thereby seeks to constrain the scientist to any single
analytical or instrumental modality. The general systems theory approach most explicitly does not so constrain the scientist. In fact, as we previously suggested, the general systems
theorist, if he had to choose among all thoughts which he
could leave his colleagues, might only ask that they consider
the suggestion that the only real truth about any phenomenon
will be that found in the concatenative nexus which is formed
between successively more specific deductive inferences and
successively more generalized inductive inferences—in short,
at the point where percept and concept collapse and where
logical and empirical, and inductive and deductive conclusions become indistinguishable.

Thus we find the general systems theorist arguing for an
end to academic parochialism and for adoption of interdisciplinary attack; arguing against simple statistical-mathematical
models (e.g., the 'shock' models of the econometricians or the
finite-state Stimulus-Response models of the behaviorist psychologists) and asking for more elegant and relevant formulations, even if these may somewhat delay the scientist's
'completion experience'; arguing against unwarrantedly deterministic paradigms and for paradigms which reflect the
inherent complexity of most phenomena of any sociobehavioral significance.

The general systems theorist can argue in these ways because his commitment is not to any discipline or school, but
to a philosophy; and his platform is one which lends him
perspective, not one which binds him to a constrained set of
techniques or which seeks to algorithmatize his searches. As
such, he is free to exercise his dedication to the subjects he
studies and to which he owes a responsibility; free from the
ritualistic defense of his predecessor's positions and freed
from the exegetical methodology which 'schoolism' implies;

and he is free from the necessity to pay homage to any academic abstractions such as Freudian psychology, functionalist anthropology, or Parsonian sociology. Nor can he draw arrogance from such abstractions; rather, he can assume only confidence from his own achievements and humility from his own failures. He is autonomous man, free to attack his own discipline and free to advance the cause of science. For when one is captured by a disciplinary dogma, one ceases to be scientist and becomes evangelist, ceases to be investigator and becomes concept-mongerer. As Coleridge once noted:

> He who begins by loving Christianity
>   better than Truth
> Will proceed by loving his own sect
>   or church better than Christianity,
> And end by loving himself better than all.[13]

NOTES AND REFERENCES

Introduction

1. From his "Reason in Science and Conduct," in: *Human Values and Natural Science* ed. Laszlo & Wilbur (New York: Gordon and Breach 1970), pp. 95–105.

2. *The Systems View of the World* (New York: George Braziller, 1972), p. vii.

3. Ibid., p. 19.

4. *Trends in General System Theory* ed. George Klir (New York: Wiley-Interscience 1972), pg. 16.

5. *Systems Theory* (Zadeh & Polak, eds.); (New York: McGraw-Hill 1969), p. vii.

6. "A Computer Approach to General System Theory," in *Trends . . .*, op. cit., pp. 98–141.

7. *Beyond Economics* (Ann Arbor: University of Michigan Press, 1968), p. 96.

1. General Systems Theory: Parameters and Promises

1. James B. Conant, *Modern Science and Modern Man* (New York: Doubleday-Anchor Books, 1952), p. 130.

2. See, for example, Chapter 2 of Kuhn's *The Structure of Scientific Revolutions* (Chicago: University of Chicago Press, 1962), particularly p. 15ff.

3. Robert Merton, *Social Theory and Social Structure* (New York: Free Press, 1968), p. 139.

4. "The Search for Simplicity," in: *The Relevance of General Systems Theory* ed. Ervin Laszlo (New York: George Braziller, 1972), pp. 13–30.

5. Anatol Rapoport, "The Search for Simplicity," in *Main Currents in Modern Thought* Vol. 28, No. 3 (Jan.–Feb., 1972), pp. 79–84.

6. This statement was made in 1951 and was reproduced in Gerald M. Weinberg's article: "A Computer Approach to General System Theory," in: *Trends in General System Theory* ed. George J. Klir (New York: Wiley-Interscience, 1972), pp. 98–141.

7. Conant, op. cit., p. 41ff.

8. Frederick Vivian, *Thinking Philosophically* (New York: Basic Books, 1969), p. 60.

9. Ibid., p. 62.

10. Errol Harris, *Hypothesis and Perception* (London: George Allen and Unwin, 1970), p. 203.

11. Kuhn suggests that 'mopping up' operations occupy the majority of scientists throughout their career and that such operations constitute *normal science*. Specifically, he suggests that: ". . . normal scientific research is directed to the articulation of those phenomena and theories that the paradigm already supplies." (Kuhn, *Structure*, op. cit., p. 24.)

12. See Merton, op. cit., Chapter 2, or van Nieuwenhuijze's: *Intelligible Fields in the Social Sciences* (The Hague: Mouton & Co., 1967), Part 1, Section 5.

13. For example, Weber once defined sociology as: ". . . a science which attempts the interpretive understanding of social action in order thereby to arrive at a causal explanation of its course and effects." From Henderson and Parson's translation of: *The Theory of Social and Economic Organization* (New York: Oxford University Press, 1947). Similarly, Maslow suggested that: "We must remember that knowledge of one's own deep nature is simultaneously knowledge of human nature in general." (From Henry Geiger's introduction to Maslow's: *The Farther Reaches of Human Nature* (New York: Viking Press, 1971). The point here is that Weber's 'interpretive' stance lends his works an idiographic quality much as Maslow's subjectivism marks his work as having a personalistic bias. In both cases, the scientific utility of the constructs is marred by idiographic and hypostatical overtones.

14. Conant, op. cit., p. 129ff.

15. Piaget, for example, includes an element of 'volunteerism' in his thesis, to complement the environmentalistic tendencies of Maslow and the biomechanical bias of Skinner. See, for example, Piaget's: *Science of Education and the Psychology of the Child* (New York: Orion Press, 1970), or his article prepared with Inhelder: "The Gaps In Empiricism," in: *Beyond Reductionism: New Perspectives in the Life Sciences* ed. Koestler & Smithies, (New York: Macmillan, 1969), pp. 118–160.

16. Ludwig von Bertalanffy, "The History and Status of General System Theory," in: *Trends in General System Theory*, op. cit., p. 25.

17. Cf., Clifford Grobstein's article: "Hierarchial Order and Neo-genesis," in *Hierarchy Theory: The Challenge of Complex Systems* ed. Howard H. Pattee (New York: George Braziller, 1973), pp. 31–47. For a more broad treatment of the concept of holistic 'personalities' see: *Beyond Reductionism: New Perspectives in the Life Sciences* (New York: Macmillan, 1969), edited by Arthur Koestler and J. R. Smithies. Especially intriguing is Koestler's article: "Beyond Atomism and Holism—The Concept of the Holon."

18. See, for example, the chapter on segmentation as a fundamental

modality for organizational evolution in Leslie A. White's: *The Evolution of Culture* (New York: McGaw-Hill, 1959).

19. This, at least would be the position of Parsons, via Comte and Durkheim, as opposed to that of Marx. Gouldner, for example, notes that for Comte, Durkheim, and Parsons, ". . . system change has to be thought of as deriving from exogenous forces, the system *model* itself not being conceived of as possessing internal sources of disequilibrium." Gouldner then notes that another line of sociological theory ". . . deriving from the Marxian tradition, stresses that the system can change due to its 'internal contradictions,' that is, endogenous forces." See Gouldner's: "Reciprocity and Autonomy in Functional Theory," in: *System, Change and Conflict* ed. Demerath and Peterson, (New York: Free Press, 1967), pp. 141–169. For an eclectic point of view on sources of change see Walter Buckley's: *Sociology and Modern System Theory* (Englewood Cliffs: Prentice-Hall, 1967), especially the survey of Parson's theories in Chapter 2.

20. Kurt Lewin, *Field Theory in the Social Sciences* (New York: Harper & Brothers, 1951), p. 151.

21. Ervin Laszlo, *The Systems View of the World* (New York: George Braziller, 1972), p. 8.

22. An excellent study of the immense philosophical contribution of Epicurus is given by Norman W. Dewitt in: *Epicurus and his Philosophy* (Cleveland: The World Publishing Company; Meridian Books, 1967). The quotation used is found on page 133.

23. Cf., Maurice Mandelbaum's: "Functionalism in Social Anthropology," in: *Philosophy, Science, and Method: Essays in Honor of Ernest Nagel* ed. Morgenbesser, et. al., (New York: St. Martin's Press, 1969), pp. 306–332.

24. Quoted by Theodosius Dobzhansky, "On Cartesian and Darwinian Aspects of Biology," in Morgenbesser, ed., *Philosophy*, op. cit., pp. 165–178. The reader may also want to see the introduction to Dubos's more recent book *So Human an Animal* (New York: Scribner's, 1968), for broader comments in the same vein.

25. Thus, for example, Walter Buckley can tell us the following: "When we say that 'the whole is more than the sum of its parts' . . . the 'more than' points to the fact of *organization*, which imparts to the aggregate characteristics that are not only *different* from, but often *not found in* the components alone; and the 'sum of the parts' must be taken to mean, not their numerical addition, but their unorganized aggregation." From: *Sociology and Modern System Theory*, op. cit., p. 42.

26. Cf., Errol Harris, *Hypothesis and Perception*, op. cit., p. 80, or Laszlo's *The Systems View of the World*, op. cit., p. 26.

27. Buckley, op. cit., p. 39.

28. "Systems and Their Informational Measures," in *Trends in General System Theory*, ed. Klir, op. cit., p. 79.

29. "The Living System: Determinism Stratified," in: *Beyond Reductionism* ed. Koestler & Smithies, op. cit., pp. 3–55.

30. At one time, such psychophysical phenomena as memory were thought to be essentially intelligible in terms similar to those used to treat other electrical or mechanical subjects. . . . i.e., via simple switching theory. Even here however, where the purportedly 'deterministic' roots of human behavior are ostensibly sunk most deeply, scientists have had to abandon such simple, physicalistic schemes in favor of vastly more complex electrochemical concepts. For a study of this gradual complexification of psychophysics, especially with respect to memory, see: D. S. Halacy, Jr., *Man and Memory: Breakthroughs in the Science of the Human Mind* (Harper & Row, 1970).

31. See, for example, the sections on deductive systems and deductive inference in R. B. Braithwaite's: *Scientific Explanation* (Cambridge: Cambridge University Press, 1953).

32. "The Uses of Mathematical Isomorphism in General System Theory," in: *Trends in General System Theory*, op. cit., pp. 42–77.

33. *Ibid.*

34. Mandelbaum, "Functionalism in Social Anthropology," in: *Philosophy, Science, and Methodology* ed. Morgenbesser, et. al., op. cit., p. 324.

35. Karl Pribham, *Conflicting Patterns of Thought* (Washington: Public Affairs Press, 1949), p. 22.

2. The Rationalizing Role of General Systems Theory: Its Ontological Implications

1. Ludwig von Bertalanffy, *Robots, Men and Minds* (New York: George Braziller, 1967), p. 94.

2. William Richard Gondin, *Prefaces to Inquiry: A Study in the Origins and Relevance of Modern Theories of Knowledge* (New York: King's Crown Press, 1941), p. 188.

3. For an explanation of the origins and extended implications of universalism and nominalism, see Karl Pribham's: *Conflicting Patterns of Thought* (Washington: Public Affairs Press, 1949), especially Chapters II and III.

4. Cf., Norman Wentworth Dewitt, *Epicurus and his Philosophy* (Cleveland: Meridian Books, 1967), p. 15ff.

5. Cf., Chapter 12 in D. S. Halacy, Jr.'s: *Man and Memory* (New York: Harper and Row, 1970), where there is an interesting discussion of experiments designed to validate the concept of transfer of learning (through biochemical means). As for a discussion of the broad implications of deep structures (a nontechnical presentation), see Noam Chomsky's: *Problems of Knowledge and Freedom* (New York: Vintage Books, 1971).

6. Thomas Kuhn, *The Structure of Scientific Revolutions* (Chicago: University of Chicago Press, 1962).

7. In this respect, see: Errol Harris, *Hypothesis and Perception* (London:

George Allen and Unwin, 1970), especially chapters 5 and 6 or James B. Conant's *Modern Science and Modern Man* (New York: Doubleday-Anchor, 1952), passim. This is a point which is frequently raised in Kuhn, op. cit.

8. Kuhn, op. cit., p. 19ff.
9. Frederick Vivian, *Thinking Philosophically* (New York: Basic Books, 1969), p. 165ff.
10. Harris, op. cit., p. 105.
11. Fernand van Steenberghen, *Epistemology* (New York: Joseph F. Wagner, 1949), p. 67.
12. Ibid., p. 69.
13. Harris, op. cit., p. 125.
14. C. Wright Mills, "Grand Theory," in: *System, Change and Conflict* ed. Demerath & Peterson (New York: Free Press, 1967), pp. 171–183.
15. Ibid., p. 172.
16. Lancelot T. Hogben, *Retreat From Reason* (New York: Random House, 1938), p. 6.
17. James K. Feibleman, *Ontology* (New York: Greenwood Press, 1968), p. 164.
18. Ibid.
19. Gondin, op. cit., p. 118.
20. Maurice Natanson, ed., *Philosophy of the Social Sciences* (New York: Random House, 1963), pg. 30f.
21. David Hume, *Enquiry Concerning Human Understanding* IV, Part I.
22. Noam Chomsky, "Some Empirical Assumptions In Modern Philosophy of Language," in: *Philosophy, Science, and Method* ed. Morgenbesser, et. al. (New York: St. Martin's Press, 1969), pp. 260–285.
23. Carl. G. Hempel, "Reduction: Ontological and Linguistic Facets," in: *Philosophy, Science and Method*, op. cit., pp. 179–199. For an expansion of these ideas, see: J. A. Shaffer's "Recent Work on the Mind-Body Problem," *American Philosophical Quarterly* (1965), 2, pp. 81–84.
24. Cf., Ervin Laszlo, *The Systems View of the World* (New York: George Braziller, 1972), passim.
25. Hempel, op. cit., p. 179.
26. Hume, op. cit., IV, Part II.
27. Inherently stochastic entities are so designated by the fact that the causal algorithms engining them allow for a given set of starting-state properties to ultimate in two or more unique ultimate states. Thus, such entities are probabilistic by their very nature. For an explanation of methods for dealing with such entities see the article by Ho and Lee: "A Bayesian Approach to Problems in Stochastic Estimation and Control," *IEEE Transactions in Automatic Control* Vol. AC-9, (1964), pp. 333–338.
28. Cf., Jean Piaget's *Science of Education and the Psychology of the Child* (New York: Orion, 1970) or Ginsburg & Opper, *Piaget's Theory of Intellectual Development* (Englewood Cliffs: Prentice-Hall, 1969). In both cases, note the emphasis on the active, creative role assigned the human

being—thus implying that future behavioral states are not simple projections available through strict inductive inference.

29. Suzanne Bachelard, *A Study of Husserl's Formal and Transcendental Logic* (Evanston: Northwestern University Press, 1968), p. xlvi.

30. Walter Buckley, "A Systems Approach to Epistemology," in: *Trends in General System Theory* ed. George J. Klir (New York: Wiley-Interscience, 1972), pp. 188–202.

31. van Steenberghen, op. cit., p. 279ff.

32. Pribham, op. cit., p. 87.

33. Conant, op. cit., p. 38.

34. Ibid., pg. 106ff.

35. Harris, op. cit., p. 202.

36. In the same sense, Paul Tillich recalls Einstein saying that the true scientist "attains that humble attitude of mind towards the grandeur of reason incarnate in existence, which, in its profoundest depths, is inaccessible to man." Quoted from Tillich's: *Theology of Culture* (New York: Oxford University Press, 1959), p. 130.

37. Taken from an article written by Walter Sullivan, appearing in the *New York Times* (March 28, 1972), quoting from a previously unpublished letter written by Einstein.

38. Ibid.

39. Werner Heisenberg, *The Physical Principles of Quantum Theory* (New York: Dover Press, 1949), p. 62.

40. See Kuhn, op. cit., especially chapter 7.

41. Cf., James D. Watson, *The Double Helix* (New York: Signet Books, 1968).

42. G. Feinberg, "On What There May Be In The World," in: *Philosophy, Science and Method*, op. cit., pp. 152–164.

43. Ibid., p. 152.

44. Ibid.

45. Arthur Eddington, *The Philosophy of Physical Science* (New York: Macmillan, 1939), p. 115.

46. As reported by W. V. Quine, in: *Ontological Relativity and Other Essays* (New York: Columbia University Press, 1969), p. 26.

3. General Systems Theory as a Counter to Scientism

1. Michael Polanyi, *The Tacit Dimension* (New York: Doubleday, 1966), p. 82ff.

2. Anatol Rapoport, "The Search for Simplicity," in: *The Relevance of General Systems Theory* ed. Ervin Laszlo (New York: George Braziller, 1972), pp. 15–30.

3. The immanence of ontological predicates, their indirect but powerful effect on scientific practices, has seldom been considered in anything but the most casual way. The work of James K. Feibleman, however, is an exception

to this (and, if anything, he is too willing to subsume empiricals and 'reifics' under ontology). At any rate, a reading of his excellent book is recommended: *Ontology* (New York: Greenwood Press, 1968).

4. Quoted from Dorwin Cartwright's introduction to Lewin's: *Field Theory in the Social Sciences* (New York: Harper & Brothers, 1951), p. ix.

5. Several books have been written which explicitly explore the problem of transferring the instruments of the natural/physical sciences to the social/behavioral sciences. Two which are especially useful in that they are 'specific' in their arguments are (a) Schoek & Wiggin's: *Scientism and Values* (New York: Van Nostrand, 1960), and; (b) Koestler and Smithies's: *Beyond Reductionism* (New York: Macmillan, 1968). Whereas the former treats the philosophy of science in respect to scientistic practices, the latter explores the necessity for going beyond traditional methodologies in the life sciences, with many implications for the human scientist.

6. i.e., equifinality is associated with the ability of certain organisms to arrive at a given end by different causal trajectories. While this phenomenon has been explored primarily by biologists with respect to embryonic developments, its implications for human and social behavior are vast. Indeed, the potential for equifinality is an implicit in Ackoff and Emery's discussion of the meaning of purposeful behavior attributed to complex systems. See their: *On Purposeful Systems* (Chicago: Aldine-Atherton, 1972).

7. For a discussion of the origin of these criteria, see Gardner Murphy's: "The Inside and Outside of Creativity," *Fields Within Fields . . . Within Fields* (New York: The World Institute, 1969), Vol. 2, No. 1, 1969, p. 7.

8. For a detailed discussion of the theoretical/empirical underpinnings of equifinality, see C. H. Waddington's: "The Theory of Evolution Today," in Beyond Reductionism, op. cit., pp. 357–395.

9. I have encountered, for example, in both Latin America and the Caribbean, what might be called *anabiotic* cultures—desiccated and empirically transparent until they have been excited (in this instance by a governmental development program). A particularly good example is the vestige of medieval Catholic economic dogma which still operates determinantly, if sub rosa, among the more removed populations in Northeast Brazil; capital accumulation has consistently been blocked by the dogmatic predicate that suggests that 'a man's wealth should not exceed the station into which he was born.' To some extent, this dogma acts in a deep-seated way to militate against any economic adventure whatsoever. With regard to the determinacy of Catholic economic dogma, see R. H. Tawney's: *Religion and the Rise of Capitalism* (London: 1924).

10. Helmut Schoek, for example, gives us an indication of this when he suggests that "Certain stocks of germs are known to outwit the researcher by selectively outbreeding his luck with resistant strains." Quoted from his Scientism and Values, op. cit., p. 136.

11. Pitirim Sorokin, *Sociocultural Causality, Space and Time* (New York: Russell & Russell, 1964), p. v.

12. Ibid., p. 236.

13. For an even stronger admission of an idiographic-interpretive preference, see: Max Weber, *The Theory of Social and Economic Organizations*, trans. Henderson & Parsons (New York: Oxford University Press, 1947): On Page 88 Weber tells us that sociology is the attempt to gain an "interpretive understanding" of social action, which for the Germanic inheritor of Verstehen means a single, universal causal system (i.e., ideational engination).

14. See his *Value in Social Theory* (London: Routledge & Kegan Paul), 1958.

15. Quoted from his: *Foundations in Sociology* (New York: Macmillan, 1939), p. 40.

16. Ludwig von Bertalanffy, *Robots, Men and Minds* (New York: George Braziller, 1967), p. 29ff.

17. For an insightful discussion of the dangers in making inferences from animal to human behavior, see Book II, Chapter XII in Arthur Koestler's: *The Act of Creation* (New York: Macmillan, 1964).

18. It is interesting to speculate what would happen to the laws of thermodynamics and such essentially metaphysical concepts as 'entropy' or the 'quark' were strict positivist standards to be obeyed by modern physics.

19. Quoted by Fritz Machlup in his "If Matter Could Talk," in: *Philosophy Science and Methodology* ed. Morgenbesser, et. al., (New York: St. Martin's Press, 1969) pp. 286–305.

20. Ibid., p. 296.

21. Ervin Laszlo, *The Systems View of the World* (New York: George Braziller, 1972), p. 79.

22. Mario Bunge, "The Metaphysics, Epistemology and Methodology of Levels," in: *Hierarchical Structures* ed. Whyte, et. al. (New York: Elsevier, 1969), pp. 17–28. Whether or not this article warrants its sweeping title I shall leave to the reader to judge.

23. Amitai Etzioni, *The Active Society* (New York: Free Press, 1968), p. 50.

24. Ibid., p. 51.

25. Paul Tillich, *The Theology of Culture* (New York: Oxford University Press, 1959), p. 91.

26. B. F. Skinner, *Beyond Freedom and Dignity* (New York: Alfred A. Knopf, 1971), p. 101.

27. Ibid.

28. von Bertalanffy, *Robots, Men and Minds*, op. cit., p. 90.

29. Ludwig von Bertalanffy, *General Systems Theory* (New York: George Braziller, 1968), p. 208.

30. A particularly interesting example of this problem occurs in the attempts to understand the human mind by using the electronic computer as an analogic referent. What happens is that we do all right up to the point where we have to suggest something about behavioral mechanisms that are

not linked (physically) to sense data. Beyond that, even the most erudite psychophysicists seem to have recourse only to speculation (or, more often, to the entirely unwarranted presumption that as we learn how to do more with computers, we shall also learn more about how conceptual-symbolic tasks take place within the 'mechanism' of the mind. At any rate, the analogy is very fruitful up to this point, and an excellent exposition is given in Dean Wooldridge's: *The Machinery of the Brain* (New York: McGraw-Hill, 1963).

31. Piaget's remark is given in the 'afterthoughts' addendum to his and Inhelder's article "The Gaps In Empiricism," in: *Beyond Reductionism, op. cit.*, p. 157ff.

32. Bacon may have set the trajectory for 'instrumental' science when he suggested (long before the Industrial Revolution) that: "I would address one admonition to all: that they consider what are the true ends of knowledge, and that they seek it neither for pleasure of the mind, or for contention, or for superiority to others . . . but for the benefit and use of life. . . . And so those twin objects, human knowledge and human power, do really meet in one." These statements were collapsed from different parts of Bacon's writings by Hans Jonas, and quoted in his: "The Practical Uses of Theory," *Social Research* (Vol. 26, No. 2, Summer, 1959), pp. 127–166.

33. Quoted from Robert Rosen's: "Hierarchical Organization of Biological Systems," in: *Hierarchical Structures, op. cit.*, p. 180.

34. Alvin Gouldner, *The Coming Crisis of Western Sociology* (New York: Basic Books, 1970), p. 28.

35. George Strauss, "Some Notes on Power Equalization," in: *The Social Science of Organizations* ed. Harold J. Leavitt, (Englewood Cliffs: Prentice-Hall, 1963), pp. 39–84.

36. Cf., Alan Ryan's *The Philosophy of the Social Sciences* (New York: Pantheon Books, 1970), p. 232. Ryan, here, gives an excellent summary of the implications of Popper's "open society."

37. At least one eminent sociologist agrees with the economists, suggesting that, ". . . they have been asking pointed and often acutely perceptive questions of other social scientists and getting vague, horatory or no answers." See Wilbert E. Moore's "The Social Framework of Economic Development," in: *Tradition, Values and Socioeconomic Development* ed. Braibanti & Spengler (Durham: Duke University Press, 1961), p. 57.

38. Jacob Marschak, "Economic Measurements for Policy and Prediction," in: *Studies in Econometric Methods* ed. Hood & Koopmans, (New Haven: Yale University Press, 1953), pg. 26.

39. Ibid., p. 14.

40. Gouldner, op. cit., p. 30.

41. Ryan, op. cit., p. 232.

42. W. I. B. Beveridge, *The Art of Scientific Investigation* (New York: Vintage Books, 1950), p. 125ff.

43. F. Kenneth Berrien, *General and Social Systems* (New Brunswick: Rutgers University Press, 1968), p. 10.

44. Katz & Kahn, *The Social Psychology of Organizations* (New York: John Wiley and Sons, 1966).

45. Ibid., p. 8. Here they suggest that "Parsons . . . more than any other person, has utilized the open-system approach for the study of social structures." There is, of course, a problem with this: that in so doing we tend to assume only exogenous factors as system determinants, and ignore intrasystem potentials for disturbance, change, or mutation. In this regard see Walter Buckley's: *Sociology and Modern Systems Theory* (Englewood Cliffs: Prentice-Hall, 1967), especially Chapter 2.

46. Jay W. Forrester, *Urban Dynamics* (Cambridge: M.I.T. Press, 1969), p. 9.

47. Quoted from: Frank George, *Models of Thinking* (London: George Allen and Unwin, 1970), p. 28.

48. Katz and Kahn, op. cit., p. 2.

49. H. Spencer, *The Study of Sociology* (London: King Press, 1875), p. 330f.

50. March and Simon, *Organizations* (New York: Wiley, 1958), p. 4.

51. Leslie A. White, *The Evolution of Culture* (New York: McGraw-Hill, 1959).

52. Gouldner, op. cit., p. 119.

53. Ernest Nagel, "Problems of Concept and Theory Formation in the Social Sciences," in: *Philosophy of the Social Sciences*, ed. Maurice Natanson (New York: Random House, 1963), pp. 188–209.

54. Ibid.

55. Quoted from: *Intelligible Fields in the Social Sciences* (The Hague: Mouton & Co., 1967), p. 254.

56. Cf., his article: "The Postulates of Science and their Implications for Sociology," in: *Philosophy of the Social Sciences*, op. cit., pp. 33–72.

57. For example, Goffman's interpretation of sociobehavioral phenomena involves such complexity and empirical transparency, that the current state of the mathematical art, and the present immaturity of game-theoretical methods, makes it almost impossible to enter any element of quantification. Cf., his: *Asylums* (New York: Doubleday, 1961), but also see Martin Shubik's: *Readings in Game Theory and Political Behavior* (New York: Doubleday, 1954) for some readily intelligible ideas on what simple game theory *can* do.

58. Cf., Chapter 2 of Arden and Astill's *Numerical Algorithms* (Reading, Massachusetts: Addison-Wesley, 1970).

59. For an explanation of his treatment of complexity, see his paper: "The Architecture of Complexity," *Proceedings of the American Philosophical Society*; 106, pp. 467–482.

4. Methodological and Instrumental Implications of General Systems Theory

1. Gardner Murphy, "The Inside and Outside of Creativity," *Fields Within Fields . . . Within Fields* (New York: The World Institute, 1969), Vol. 2, No. 1, pp. 7–9.

2. For example, with respect to the quotation from Gardner Murphy, it was Watson and Crick who 'deduced' the necessity for the DNA structure to be the double helix, largely because the helix was the logically preferable structure. Rosa Franklin kept her defraction X rays out of the hands of the theorists until after the formulation had been conceptually completed. In this case, then, despite Murphy's allocation of the discovery of the DNA structure to empirical science, the discoverers were actually operating very close to what we might call the Neoplatonic modality, always within the confines of the hypothetico-deductive modality. The key here is this: the empirical validation of the theory was accomplished only after the fact of discovery, as the empirical proofs were not available to Watson and Crick during the discovery process itself. For a complete history of this remarkable scientific achievement, see: James D. Watson, *The Double Helix* (New York: Signet Books, 1968).

3. As we have previously sought to explain, there is a critical difference between *inherently* stochastic entities and *effectively* stochastic entities, etc. The former are those which will be expected to remain stochastic, irrespective of the amount of analytical time or resources we expend, due to their inherent nature, whereas the latter are expected to yield to analysis, ultimately reaching determinacy. It is not too much to suggest that the normal concept of science envisions all phenomena in the latter category, though certain of the more advanced theories in quantum physics, for example, are giving explicit credence to the possibility of certain phenomena being inherently nondeterministic but nevertheless proper subjects for description and analysis. In the human sciences, we have an implicit emphasis on the latter, which at least partially accounts for our humility about the ultimate ends of our research into social and behavioral subjects. In short, Skinner and the behaviorists and certain small-group theorists in sociology have been primarily concerned with effective indeterminacy, whereas men such as Sorokin and Weber and Vickers might be said to have been dealing, all along, with inherently stochastic phenomena without the express ambition of ever reducing them to deterministic entities. A good contrast, then, might be between a reading of Skinner's *Beyond Freedom and Dignity* (New York: Knopf, 1971) and, say, Sorokin's *Sociocultural Space, Causality and Time* (New York: Russell & Russell, 1964).

4. Cf., Werner Heisenberg's *The Physical Principles of Quantum Theory* trans. Eckart & Hoyt, (New York: Dover Publications, 1949), especially Chapter IV.

5. As an example of an inherently indeterminate system, Helmut Schoek has noted that: "Certain stocks of germs are known to outwit the antibiotics researcher by selectively outbreeding his luck with resistant strains." Quoted from his and Wiggin's *Scientism and Values* (New York: Van Nostrand Reinhold, 1960).

6. For a concise, mathematically predicated explanation of the Bayesian-driven learning process, see: K. S. Fu, "Learning System Theory," in *System Theory* ed. Zadeh & Polak, (New York: McGraw-Hill, 1969), pp. 425–463. Or, for a simpler, less mathematically sophisticated explanation, see Chapter 6 of Frank George's *Models of Thinking* (London: George Allen and Unwin, 1970).

7. However, as we have so many times pointed out, it is very doubtful that the really significant discoveries of the natural and physical sciences owe much to the process where empirical data collection precedes hypothesis-building, or where no deductive elements enter. But W. I. B. Beveridge offers another element which had seemed to contribute greatly at various times to significant discovery—the elements of chance and intuition and imagination, etc. See, for example, Chapters 3, 5, and 6 of his *The Art of Scientific Investigation* (New York: Vintage Book, 1957).

8. Some of the more obvious would be, for example, the degree of homogeneity among the properties of the system, the degree of symmetry in their distribution, etc.

9. Cf., Kahn and Weiner, *The Year 2000: A Framework for Speculation* (New York: Macmillan, 1967).

10. Fred Emery, "The Next Thirty Years: Concepts, Methods and Anticipations," *Human Relations* 20 (1967), p. 199–235.

11. See, for example, Newell, Shaw, and Simon's: "Elements in a Theory of Human Problem-Solving," *Psychological Review* Vol. 65 (1958), pp. 151–166.

12. For a discussion of the difference between association and bisociation (where bisociation is a combination of conceptual entities on a very complex and creative level), see Arthur Keostler's *The Ghost in the Machine* (New York: Macmillan, 1967), pp. 182ff.

13. *From Aids to Reflection: Moral and Religious Aphorisms*, XXV.

Index

205

INDEX

INDEX